Workshop on Materials and Engineering in Aeronautics (MEA2020)

IOP Conference Series: Materials Science and Engineering
Volume 1027

Moscow, Russia
16 - 17 October 2020

Editors:

Evgeny V. Nikulchev
Oleg N. Korsun

ISBN: 978-1-7138-3327-7
ISSN: 1757-8981

Printed from e-media with permission by:

Curran Associates, Inc.
57 Morehouse Lane
Red Hook, NY 12571

Some format issues inherent in the e-media version may also appear in this print version.

Institute of Physics
Dirac House, Temple Back
Bristol BS1 6BE UK

Phone: 44 1 17 929 7481
Fax: 44 1 17 920 0979

techtracking@iop.org

Additional copies of this publication are available from:

Curran Associates, Inc.
57 Morehouse Lane
Red Hook, NY 12571 USA
Phone: 845-758-0400
Fax: 845-758-2633
Email: curran@proceedings.com
Web: www.proceedings.com

TABLE OF CONTENTS

APPROBATION OF THE DATA CLASSIFICATION METHOD BASED ON THE SVM ALGORITHM AND THE K NEAREST NEIGHBORS ALGORITHM 1
L A Demidova, Yu S Sokolova

HYBRID EVOLUTIONARY ALGORITHM IN THE OPTIMIZATION PROBLEM OF THE SHOOTING SYSTEM PARAMETERS 10
L A Demidova, N A Petrova

SYNERGY OF QFD AND FMEA METHODS TO IMPROVE WORKPLACES IN AIRCRAFT WIRE MANUFACTURING 17
A Ya Dmitriev, T A Mitroshkina, I V Russkikh

THE PREDICTIVE DISPLAY AS A MEANS OF SUPPRESSING THE NEGATIVE EFFECTS OF WAKE VORTEX ENCOUNTERS 23
A V Efremov, M S Tiaglik, T V Voronka

NUMERICAL SIMULATION OF SPLASHING LANDING GEAR ON AN AMPHIBIOUS AIRCRAFT AIR CUSHION 29
A V Fevralskikh

TECHNOLOGY FOR AERODYNAMIC WAKE CHARACTERISTICS MEASUREMENT BEHIND AN AIRCRAFT ON A RAIL-GUIDED TRACK 36
I G Golovnev, V V Vyshinsky, O V Sokolov, K V Lapshin, P A Mukhachev

ABOUT THE EQUILIBRIUM OF AIRSHIP IN THE PARKING LOT 41
V T Grumondz, N I Morozov, Moung Htang Om

HOVERING CONTROL ALGORITHM VALIDATION FOR A MOBILE PLATFORM USING AN EXPERIMENTAL TEST BENCH 46
S Jatsun, O Emelyanova, B Lushnikov, A S Martinez Leon, L M Mosquera Morocho, A Pechurin, C A Nolivos Sarmiento

NONLINEAR STOCHASTIC ESTIMATION OF ALTITUDE AND VERTICAL FLIGHT SPEED IN REAL TIME 53
V P Kharkov, O N Korsun, M N Pravidlo, Moung Htang Om

LAYERED COMPOSITE FOR A GEOPOLYGON MIRE 59
A P Kondratov, G N Zhuravleva, Y M Syltanova

DETECTION OF DYNAMIC ERRORS IN AIRCRAFT FLIGHT DATA 63
O N Korsun, S V Nikolaev, Moung Htang Om

SPEECH SPECTRUM ANALYSES FOR ESTIMATING OPERATOR FUNCTIONALITY 70
O N Korsun, V N Yurko, S A Beyseev, A S Naukenova, A K Tulekbaeva

DIGITAL RECRUITMENT OF INNOVATIVE PROJECT TEAMS' LEADERS BASED ON GENETIC ALGORITHMS 76
E E Kovshov, V S Kuvshinnikov, L E Osipenko

HIERARCHY ANALYSIS FOR MONITORING THE UNIFORMITY OF MEASUREMENTS SYSTEM 84
I V Krasavin, A V Yashin, A K Tulekbaeva

MATHEMATICAL FOUNDATION OF SYSTEM DESIGN REGULATORY CONTROLS BASED ON ARP4754A 91
Vladimir Kulabukhov

TO THE QUESTION OF CALCULATING THE RUN DISTANCE OF THE LIGHTWEIGHT WIG PLAN AT THE STAGE OF PRELIMINARY DESIGN 97
A N Luchkov

DEVELOPMENT OF THE MATHEMATICAL MODEL OF CATALYTIC OXIDATION-REDUCTION REACTION KINETICS FOR WASTEWATER TREATMENT TECHNOLOGY USING VORTEX LAYER DEVICES 103
P E Lukashev, M I Sidorov, M E Stavrovskiy, S A Lazarev, A V Ragutkin, E I Korolev, N Sh Gazanova

PRODUCTION EXPERT SYSTEM TO DETERMINE AIRCRAFT TAKE-OFF AND LANDING PERFORMANCE 115
A V Melnichuk, V A Nesterov, V A Sudakov, K I Sypalo

ANALYSIS OF POST BUCKLING STATE OF COMPOSITE RECTANGULAR PANEL WITH ASYMMETRIC STRUCTURE ON ELASTIC BASE CAUSED BIAXIAL COMPRESSION 118
O Mitrofanov, M Klesareva

POST-BUCKLING BEHAVIOR ESTIMATION OF COMPOSITE PANELS OF ASYMMETRIC STRUCTURE WITH ALL-BOUNDARY RIGID SUPPORT WITH SHEAR FLOWS ACTION 124
O Mitrofanov, M Osman

BAYESIAN RECOGNITION OF A MOVING OBJECT INFORMATION-MEASURING SYSTEM STATE: A PRIORI INFORMATION WEIGHT CORRECTION 131
S M Muzhichek, G G Sebryakov, V I Pavlov, S V Tolstych, M B Kenzhekhanova

HYDRAULIC LOSSES OPTIMIZATION METHODS USED IN AIR CONDITIONING SYSTEM VALVE DESIGN FOR DIFFERENT CLOSURE ANGLES 135
K S Napreenko, A V Lamtyugina

CONTROL SYNTHESIS FOR TASKS WITH A GIVEN OUTPUT FUNCTION DECENTRALIZED SISO-SYSTEMS 142
E Nikulchev

SURFACE MICROGEOMETRY MONITORING OF LARGE-SIZED AIRCRAFT ELEMENTS 146
V I Pronyakin, V L Skrypka, B T Abykanova

CRITERION AND PRINCIPLES OF EASILY-MODIFIED PROGRAM 151
A Prutzkow

FUZZY JOINT OPTIMIZATION OF THE DESIGN AND OPERATING MODES OF THE INDUCTION HEATER 158
N G Rogachev, G N Rogachev

ADAPTIVE IMAGING SYSTEM FOR ELECTROMAGNETIC SCATTERING FIELD OF AIRCRAFT 165
A T Tarlanov, Z M Kurbanismailov

SYNTHETIC PSEUDO-RANDOM SEQUENCE 175
A A Tsatsin

IMPROVE THE INTEGRITY TESTING PROCESS BASED ON QFD, FMEA AND TRIZ 178
M V Vysotskaya, A Ya Dmitriev

INVESTIGATION OF CHANGES IN THE OPERATOR'S STATE BY ANALYZING THE CHARACTERISTICS OF BLINKING 186
V N Yurko, A B Uali, A S Naukenova

IMPROVING THE QUALITY OF FDM 3D PRINTING OF UAV AND AIRCRAFT PARTS AND ASSEMBLIES BY PARAMETRIC SOFTWARE CHANGES 190
R S Zagidullin, N I Zezin, N V Rodionov

Author Index

IOP Conf. Series: Materials Science and Engineering 1027 (2021) 011001 doi:10.1088/1757-899X/1027/1/011001

PREFACE

2020 Workshop on Materials and Engineering in Aeronautics (MEA2020) was held 16-17 October 2020 in Moscow Aviation Institute (National Research University) in living format using a window between the pandemia waves. This Workshop was held 16 times from 2001 to 2019. Each was attended by over 20 participants who submitted their reports for publishing. The Scientific Committee in 2020 includes:

Demidova L.A., professor, MIREA – Russian Technological University, Moscow, Russia;
Efremov A.V., professor, Moscow aviation institute, Moscow, Russia;
Karpenko A.P., professor, Bauman Moscow State Technical University, Moscow, Russia;
Korsun O.N., professor, State Research Institute of Aviation Systems, Moscow, Russia;
Evdokimenkov V.N., professor, Moscow aviation institute, Moscow, Russia;
Nikulchev E.V., professor, MIREA – Russian Technological University, Moscow, Russia;
Poplavsky B.K., professor, Gromov Flight Research Institute , Zhukovsky, Moscow Region, Russia;
Porshnev S.V., professor, Ural Federal University, Ekaterinburg, Russia;
Prutzkow A,V., professor, Ryazan State Radio Engeneering University, Ryazan, Russia;
Rogachev G.N., professor, Samara State Technical University, Samara, Russia;
Sebryakov G.G. a.m. RAS, professor, State Research Institute of Aviation Systems, Moscow, Russia;
Sizykh V.N., professor, Irkutsk State Transport University, Irkutsk, Russia;
Tulekbaeva A.K., professor, M. Auezov South Kazakhstan University, Shymkent, Kazakhstan;
Zhelannikov A.I., professor, Zhukovsky Central Aerohydrodynamic Institute , Zhukovsky, Moscow Region, Russia.

The participants represented mostly the leading technical universities of Russia. This year the international involvement is marked by the increased number of participants from Republic of Kazachstan. The (MEA2020) Workshop Proceedings include 33 papers, and the scope is actually huge. It covers materials & structure development, aircraft, airships and WIG plans design, pilot interface and pilot state monitoring, studies on aircraft wake vortex, system and control theory, statistical classification. This year we include more papers dealing with quality management and business processes control. In the year 2020 the Scientific Committee's confirms its traditional position not to limit the creativity of the participants. As a result all the papers are unlike one another and many of them contain original ideas and approaches. That is why our advice to a reader is not to judge the Proceedings by one or two papers but look through the most part of them.

On behalf of the Workshop Scientific Committee
the scientific editors of MEA2020 Proceedings:

Evgeny V. Nikulchev, Dr., Professor of MIREA – Russian Technological University, nikulchev@mail.ru

Oleg N. Korsun, Dr., Prof., Head of the laboratory, State Research Institute of Aviation Systems, marmottoduo@yandex.ru

IOP Conf. Series: Materials Science and Engineering 1027 (2021) 011002 doi:10.1088/1757-899X/1027/1/011002

Workshop MEA 2020 peer review declaration

All papers published in this volume of *IOP Conference Series: Materials Science and Engineering* have been peer reviewed through processes administered by the Editors. Reviews were conducted by expert referees to the professional and scientific standards expected of a proceedings journal published by IOP Publishing.

Type of peer review:

Double-blind , that is author and reviewer identities are hidden to each other.

Criteria used by Reviewers when accepting/declining papers:

technical, quality and presentation criteria according to IOP Conference Series peer review policy. The most part (75%) of the accepted papers were resubmitted after two or three revisions.

Conference submission management system:

through editors and reviewers personal e-mail, the special submission management systems were not employed because the total number of Workshop papers is not great.

Number of submissions received: 51 .

Number of submissions sent for review: 44 .

Number of submissions accepted: 33 .

Acceptance Rate (Number of Submissions Accepted / Number of Submissions Received X 100): 65% .

Average number of reviews per paper: 2 .

Total number of reviewers involved: 21 .

Any additional info on review process : the editors paid the especial attention on the stylistics of the papers because for all the authors English is a foreign language .

Contact person for queries:

Evgeny V. Nikulchev, Dr., Professor of MIREA – Russian Technological University, nikulchev@mail.ru

Oleg N. Korsun, Dr., Prof., Head of the laboratory, State Research Institute of Aviation Systems, marmottoduo@yandex.ru

The scientific editors of MEA2020 Proceedings:

Evgeny V. Nikulchev
Oleg N. Korsun

IOP Conf. Series: Materials Science and Engineering 1027 (2021) 012001 doi:10.1088/1757-899X/1027/1/012001

Approbation of the data classification method based on the SVM algorithm and the k nearest neighbors algorithm

L A Demidova[1] and Yu S Sokolova[2]

[1] MIREA – Russian Technological University, Moscow 119571, Russia
[2] Ryazan State Radio Engineering University named after V.F. Utkina, Ryazan, Russia

demidova.liliya@gmail.com

Abstract. The aim of the work is to test the classification method based on the combined use of SVM and *k*NN classifiers. At the first stage of the classification method, on the basis of the initial training dataset U, SVM classifier is developed and the width of the Ω area is determined. This Ω area contains all objects incorrectly classified by SVM classifier, which form together with correctly classified objects that have fallen into the Ω area and the corresponding object class labels from the Ω area a new dataset *G*. At the second stage of the classification method, the *k*NN classifier developed on the basis of information about the objects of the set $U\backslash G$ is applied to all objects of the dataset *G* from the Ω area. The values of the parameters of the *k*NN classifier are determined experimentally in such a way as to ensure the highest possible accuracy of object classification. Since the correctly classified objects can also enter the Ω area formed in the above way, the condition for the applicability of the proposed method is a general improvement in the classification quality. The presented results of experimental studies confirm the effectiveness of the proposed method in the problem of classifying complex multidimensional data.

1. Introduction

Currently, for a variety of applied problems using data of various nature and volumes, dozens of algorithms and methods of classification and their modifications, among which the most famous are, in particular, linear and logistic regression [1], Bayesian classifier [1], decision trees [1–3], decision rules [4], neural networks [5], the nearest neighbors algorithm (*k*NN algorithm, *k* Nearest Neighbors Algorithm) [6–8], support vector algorithm (SVM algorithm, Support Vector Machine Algorithm) etc. [9–11], have been developed.

In recent years, SVM algorithm [9–11] has been successfully used to solve many classification problems. SVM algorithm provides the construction of a binary SVM classifier by using a special function called the kernel function to put the vectors of characteristics of classified objects into a space of a higher dimension and search in this space for a hyperplane with a maximum gap which separates objects with different class affiliation. Despite the fact that the generalization ability of SVM algorithm is better than that of other algorithms and classification methods, there are difficulties in its application associated with the choice of the type of kernel function, the values of the kernel function parameters and the value of the regularization parameter that affect the quality of data classification. Most of the erroneously classified objects fall inside the strip separating the classes. In this regard, it is advisable to develop methods to improve the accuracy of classification solutions by reducing the number of errors within the dividing strip.

One of the modern approaches to solving the problem of increasing the accuracy of classification solutions involves ensemble in one way or another different classifiers in order to obtain a final classification solution of a higher quality. Ensembles of classifiers make fewer classification errors than each member of the ensemble individually. It is obvious that ensemble can be quite successfully applied in the case of using SVM classifier. A herewith, it is possible both to create an ensemble consisting only of some SVM classifiers [12], and to combine in the ensemble along with SVM classifier of some other classifier, which is fundamentally different from SVM classifier in terms of the instrumentation used in it [13, 14].

Since the development of a SVM classifier is associated with significant time costs for determining the optimal type of the kernel function, the values of the kernel function parameters and the value of the regularization parameter, one of important requirements which can be presented to the new classifier introduced into the ensemble, along with the requirement to ensure high accuracy classification, is the requirement of insignificance of the time spent on developing the classifier.

As such a classifier, in particular, *k*NN classifier can be used. This classifier will improve the overall accuracy of data classification while ensuring a slight increase in time costs under certain conditions described below. *k*NN classifier is the simplest metric classifier based on the assessment of the similarity of some object with other k objects. In this case, the classified object belongs to the class to which most of its *k* nearest neighbors-objects belong [6–8, 13, 14].

Currently, a number of approaches which implement the joint use of SVM and *k*NN classifiers are known. So, in [13], it is proposed to use a local SVM classifier to classify an object mistakenly classified by *k*NN classifier, using data on the nearest neighbors of this object when developing an SVM classifier. In [14], it is proposed to use in the development of *k*NN classifier, which should clarify the class belonging of objects within the dividing strip, information about the support vectors of SVM classifier. Nevertheless, all the proposed approaches are still far from perfect, and their effectiveness is obvious only when solving classification problems with a certain type of data structure. It is obvious, in particular, that the development of local SVM classifiers is associated with additional time costs both for determining the optimal number of neighbors of the classified object, and for the direct development of SVM classifier, and the use of information about SVM classifier support vectors in the development of *k*NN classifier requires preliminary thorough verification of the objectivity of their definition.

2. Theoretical research

SVM classifier provides an acceptable quality of classification on vast majority of complex multidimensional datasets [9–11]. A herewith, the analysis of the location of objects incorrectly classified by SVM classifier showed that most of them fall inside the strip separating the classes. It is proposed to refine the classification solution for objects inside the dividing strip using the proposed classification method based on the joint use of SVM and kNN classifiers [15].

The proposed method of data classification can be implemented as follows.

Stage 1. Development of SVM classifier with determination of the possibility of developing kNN classifier.

1.1. The development of SVM classifiers with the subsequent selection of the best of them in the sense of ensuring the highest classification accuracy on the dataset U is carried out on the basis of randomly generated training and test samples. The type of the kernel function, the values of the kernel parameters, and the value of the regularization parameter specified during the development of SVM classifier determine the hyperplane dividing objects into two classes with labels "–1" and "+1".

1.2. Determination of areas Ω^- and Ω^+, containing all erroneously classified objects that appeared in the sets Z^- and Z^+, respectively; d_{Ω^-} as the width of the Ω^- area; d_{Ω^+} as the width of the Ω^+ area; N_{Ω^-} as the number of objects in the Ω^- area; N_{Ω^+} as the number of objects in the Ω^+ area. Formation on the basis of the Ω^- and Ω^+ areas the final Ω area, which includes all incorrectly classified objects, forming, together with correctly classified objects that fall into the Ω area, and the corresponding object class labels from the Ω area, the dataset in which each tuple $< z_i, y_i >$ contains information about an object z_i from Ω area, and the corresponding class label $y_i \in Y = \{-1;+1\}$ for the object z_i.

It is possible to use two variants for the formation of the Ω area:

- asymmetric Ω area with respect to the dividing hyperplane: $\Omega = \Omega^- \cup \Omega^+$;
- symmetric Ω area with respect to the dividing hyperplane, containing objects which are relative to the dividing hyperplane at the distance not exceeding Δ: $\Delta = \max\{d_{\Omega^-}, d_{\Omega^+}\}$.

1.3. Formation of the dataset $W = U \setminus G$ by removing tuples of the dataset G from the dataset U. The set W will only consist of those tuples of the set U, the class of objects for which SVM classifier has been defined correctly. The objects of this set will subsequently be used to develop *k*NN classifier. Since, in addition to the inorrectly classified objects, there can be a certain number of objects classified correctly in the Ω area, it is possible that the number of tuples in the set W will be significantly less than in the set U and they will not be enough for the subsequent development of *k*NN classifier.

Stage 2. Development of *k*NN classifier.

2.1. Development of *k*NN classifiers on the basis of the dataset $W = U \setminus G$, which establish the class affiliation of all objects of the Ω area for different values of the number k of nearest neighbors from the set W using different voting methods and different methods for assessing the proximity between objects. Choosing the best *k*NN classifier in the sense of ensuring the highest classification accuracy of all objects of the Ω area and fixing the values of the parameters of the best *k*NN classifier: a variant of the considered Ω area, the used method for assessing the proximity between objects, the voting method and the optimal number of neighbors.

2.2. Comparison of the quality of the final data classification using the best *k*NN classifier with the classification quality obtained after the development of SVM classifier in order to identify the feasibility of using the classifier formed in this way to determine the class belonging of new objects, while the feasibility of using is determined by improving the quality indicators of the classification of objects from the set Z.

If the expediency of using developed classifier for the classification of new objects is revealed, it can be performed in accordance with the following sequence of steps:

- dividing new objects into two classes using developed SVM classifier;
- selecting from the new objects those that are inside the Ω area formed in the step 1.2, and refining classification solution for these objects using *k*NN classifier (with the method for assessing the proximity between objects, the voting method and the number of neighbors defined in the step 2.1 at which the quality of classification of objects from the set Z is improved).

When implementing the proposed classification method, the development of *k*NN classifier is carried out for different values of the number k of nearest neighbors. In the case of binary classification, when using a simple unweighted vote.

When developing *k*NN classifier, one will have to deal with not the entire initial dataset U, but with a set of significantly lower cardinality, containing only information about the objects that fell into the Ω area, as well as about their class affiliation. The use of a new additional toolkit such as *k*NN classifier, which uses the principles of data mining that are different from the principles laid down in SVM classifier, will in some cases improve the overall accuracy of data classification.

3. Experimental results

For the visual presentation of the results of applying the proposed method in the problem of binary classification, the *Demo* dataset, including 115 objects with two characteristics ($q = 2$) was considered. A herewith, the class of belonging of 95 objects ($s = 95$) included in the training dataset U was predetermined, and the class belonging of 20 objects included in the set V had to be established.

Figure 1 shows the location of objects from the *Demo* dataset in the D-2 space with the division of the objects set Z into the set $Z^{\times}$, the objects of which are marked with the "cross" marker and belong to the first class, and the set Z^*, the objects of which are marked with the "asterisk" marker and belong to the second class ($Z = Z^\times \cup Z^*$). Objects of the set V with unknown class are marked with the "triangle" marker.

During the implementation of the first stage of the proposed classification method in the development of SVM classifier using a radial basis function with default parameters ($\sigma = 1$ and $C = 1$), the curve dividing the classes on the randomly generated training and test sets was built (Figure 2). The size of the test set was 20% of the size of the dataset U. Objects of each class included in the test set are marked with markers smaller than the objects which form the training set. The objects, the belonging class of which must be determined (marked with the "triangle" marker in Figure 1), are not shown in Figure 2. Support vectors defined by SVM classifier (in the amount of 24 pieces) are additionally marked with the "circle" marker. Objects, the class of which is determined incorrectly by SVM classifier, are designated by numbers 1–4.

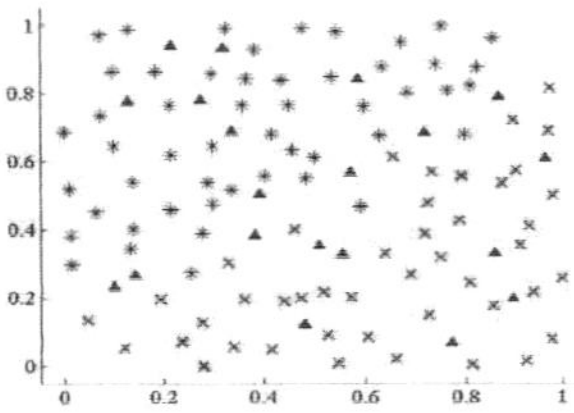

Figure 1. Presentation of the *Demo* dataset in the D-2 space.

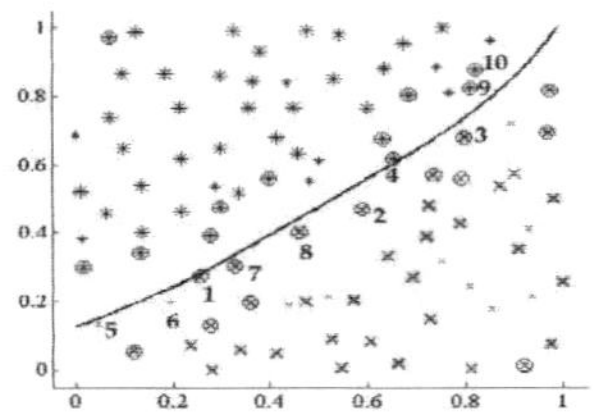

Figure 2. Splitting the *Demo* dataset into classes.

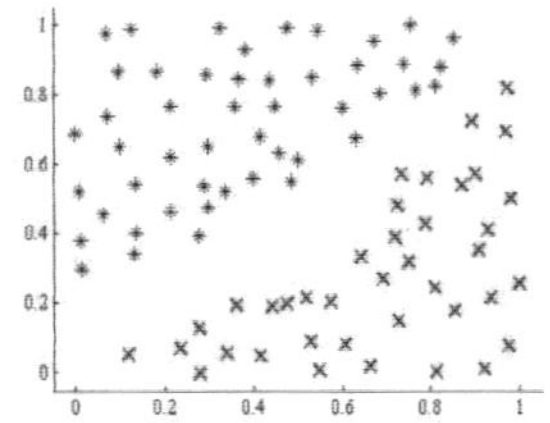

Figure 3. Training dataset without Ω area objects

As can be seen from Figure 2, three objects from the set Z^*, namely objects with numbers 1, 2, and 3, were incorrectly assigned by SVM classifier to the set $Z^\times$ associated with the first class, and the object with number 4 belonging to the set $Z^\times$ was incorrectly assigned to the set Z^* associated with the second class. Thus, SVM classifier made 4 errors on the training set, and when classifying the objects included in the test set, SVM classifier did not make mistakes. Moreover, all 4 incorrectly classified objects are defined by SVM classifier as support vectors.

Assessment of the classification quality showed that the value of the F-measure is 96%.

Before the implementation of the second stage of the proposed classification method, for each class, the following was preliminarily determined: areas Ω^* and $\Omega^\times$, containing all incorrectly classified objects which appeared in the sets Z^* and $Z^\times$ respectively; d_{Ω^*} is the Ω^*-area width; $d_{\Omega^\times}$ is the $\Omega^\times$ area width; N_{Ω^*} is the number of objects in the Ω^* area; $N_{\Omega^\times}$ is the number of objects in the $\Omega^\times$ area; the resulting Ω area containing all incorrectly classified objects (a herewith, a certain number of objects classified correctly can be located in the Ω area).

It should be noted that for the *Demo* dataset, the designations adopted in the theoretical part for the sets Z^- and Z^+, Ω^- and Ω^+ areas, the width of the Ω^- and Ω^+ areas and, the number of objects N_{Ω^-} and $\Omega^\times$ are replaced for better visualization with the designations Z^* and $Z^\times$, Ω^* and $\Omega^\times$, d_{Ω^*} and $d_{\Omega^\times}$, N_{Ω^*} and $N_{\Omega^\times}$ respectively.

Refinement of the classification results for objects included in the Ω area containing all incorrectly classified objects was carried out using the kNN algorithm. We used two variants for the Ω area:

- asymmetric with respect to the dividing curve Ω area: $\Omega = \Omega^* \cup \Omega^\times$;
- symmetric with respect to the dividing curve Ω area, containing objects located relative to the dividing curve at a distance not exceeding Δ: $\Delta = \max\{d_{\Omega^*}, d_{\Omega^\times}\}$.

Variant 1. When using the asymmetric Ω area on the basis of developed SVM classifier, it was obtained that $d_{\Omega^*} = 0{,}079$, $N_{\Omega^*} = 1$, $d_{\Omega^\times} = 0{,}503$, $N_{\Omega^\times} = 7$. There are 8 objects inside the Ω area (these are objects with numbers 1 – 8 in Figure 2).

As a result of removing from the set U the tuples of the set G containing information about the objects in the Ω area and the corresponding to them class labels, the dataset $W = U \setminus G$, consisting only of tuples, for which the class belonging was correctly determined by SVM classifier (Figure 3), was obtained.

Based on the information about the class belonging of the objects of the dataset $W = U \setminus G$ for the objects of the Ω area, the development of *k*NN classifier with the number of neighbors varying from 1 to 51 (with a step of 2) and various methods for assessing the proximity of objects was carried out. When assessing the quality of *k*NN classification, we compared the labels of the classes of objects in the Ω area, determined using *k*NN classifier, and the original labels of the classes of objects in the Ω area, fixed in the set U. *k*NN classifier with the number of neighbors equal to 7 was determined as the best one. A herewith, the estimation of the proximity between the object and its neighbors was carried out using the Euclidean measure with unweighted voting. As a result, the number of errors in the classification of objects in the Ω area was reduced from 4 to 2, that allow to increase the value of the *F*-measure to 98% (which is 2% higher than the value of the *F*-measure of SVM classifier).

Variant 2. When using the symmetric Ω area based on developed SVM classifier, it was obtained that $d_\Omega = 1{,}005$, $N_\Omega = 10$. 10 objects got inside the Ω area (these are objects with numbers 1 – 10 in Figure 2). As a result of removing from the set U the tuples of the set G containing information about the objects in the Ω area and the corresponding to them class labels, the dataset $W = U \setminus G$, which consists only of tuples of the set G, for which the class membership was correctly determined by SVM classifier, was obtained. On the basis of this dataset, as in option 1, the development of *k*NN classifier with the number of neighbors varying from 1 to 51 (with the step of 2) and various methods for assessing the proximity of objects, was carried out. When assessing the quality of *k*NN classification, we compared the labels of the classes of objects in the Ω area, determined using the *k*NN classifier, and the original labels of the classes of objects in the Ω area, fixed in the set U. *k*NN classifier with the number of neighbors equal to 3 was determined as the best one. The degree of proximity between the object and its neighbors was determined using the Euclidean metric with unweighted voting. As a result, the number of classification errors for objects in the Ω area was reduced from 4 to 3, that allow to increase the value of the *F*-measure to 97.03% (which is 1.03% higher than the value of the *F*-measure of SVM classifier).

Since in this case, with the combined use of SVM and *k*NN classifiers, the classification quality of the objects belonging to the Ω area was improved, the proposed method was applied to classify new objects (marked in Figure 1 with the “triangle” marker). A herewith, when refining the classification results using kNN algorithm for objects included in the Ω area, the classification quality was improved for each variant of the Ω area (that is, for both symmetric and asymmetric Ω areas). However, since for the dataset U a greater improvement was achieved when considering the asymmetric Ω area, then when classifying new objects belonging to the Ω area, its asymmetric version was considered.

As a result of applying developed SVM classifier to new objects, they were divided into sets Z^* and $Z^\times$ corresponding to the classes labeled “asterisk” and “cross”. In Figure 4, the new objects are marked with the ”triangle” marker. Objects from the training and test sets which fall into the asymmetric Ω area are not shown in Figure 4. Out of 20 new objects, two objects fell into the asymmetric Ω area: in Figure 4 they are designated by numbers 11 and 12, while SVM classifier identified object with number 11 in the set Z^* corresponding to the class with the label “asterisk”, and object with number 12 in set $Z^\times$ corresponding to the class with the “cross” label.

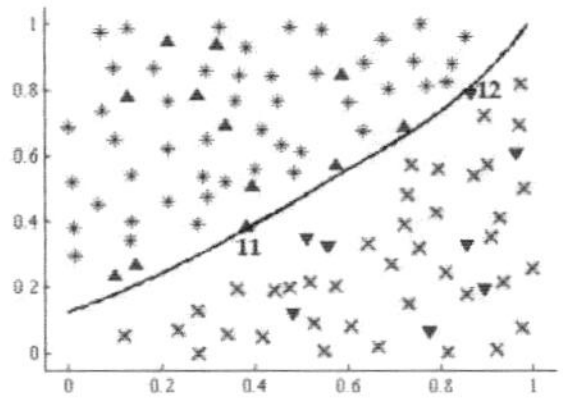

Figure 4. SVM classification results (without Ω area).

For new objects inside the asymmetric Ω area (that is, for objects numbered 11 and 12), the classification results were refined using the *k*NN classifier. The values of the parameters of *k*NN classifier, which ensure the highest possible accuracy of object classification, were determined experimentally: it was found that the best quality of classification is provided when considering the asymmetric Ω area, in which the degree of proximity between objects is determined using the Euclidean metric with unweighted voting with the optimal number of neighbors, equal to 7.

Application of *k*NN classifier to object numbered 11 did not change its class membership, and object numbered 12 was assigned by *k*NN classifier to the set Z^* corresponding to the class with the "asterisk" label, i.e. *k*NN classifier changed the result of SVM classification.

The feasibility of using the proposed classification method was also confirmed on real datasets taken from the Statlog project and the UCI Machine Learning Repository, traditionally used to test the developed machine learning algorithms. All datasets corresponds to a binary classification case.

During the experiments, it turned out that the application of the proposed classification method to some datasets turned out to be successful, but not to others, which can be explained by the already mentioned features of the formation of the Ω area or grouping of objects of the set Z within it.

Table 1 shows the values of the parameters of the classifiers and the values of the classification quality indicators for the data sets in the case of the successful application of the proposed classification method.

For each set, Table 1 presents the values of the classifier parameters and the values of the classification quality indicators for SVM classifier developed on randomly generated training and test samples using the radial basis function, which turned out to be the best (in comparison with other kernel functions) in terms of ensuring a high quality classification data. When developing SVM classifiers, the default values of the kernel parameters and the value of the regularization parameter ($\sigma = 1$ and $C = 1$) were used. The size of the test sample is 20% of the size of the dataset U.

Column 3 of Table 1 shows what type of classifier was used for the corresponding dataset: "*SVM*" record means that SVM classifier was developed; the "+ aSim" record means that after the development of SVM classifier, the classification results were refined using *k*NN classifier for objects that fell into an area asymmetric with respect to the dividing hyperplane (variant 1); the "+ Sim"record means that after the development of SVM classifier, the classification results were refined using *k*NN classifier for objects that fell into the Ω area symmetric relative to the dividing hyperplane (variant 2). The value of the width of the Ω area (d_Ω) and the number of objects inside it (N_Ω) are shown in columns 4 and 5, respectively.

Table 2 shows that the use of the proposed classification method made it possible to increase on average the value of the *F*-measure indicator by 2.5%, and the value of the classification accuracy indicator by 2.1% compared to the corresponding values obtained only with the use of SVM classifier. Moreover, the maximum increase in the *F*-measure indicator was 7.89%, and the classification accuracy indicator was 5.25%.

After the development of SVM classifier, the assessment of the number of objects which fell inside the dividing strip was made. For the training sets under consideration, in all cases presented in Table 2, all 100% of the erroneously classified objects were inside the dividing line. In addition, for all sets, the possibility of using *k*NN classifier was analyzed: the width of the Ω area, into which all objects from the training data set mistakenly classified by SVM classifier, fall, and the number of objects inside it, were estimated. In all experiments, information on which is presented in Table 1, it turned out that the

Ω area is located inside the dividing strip. As a result of removing from each set U (Table 1) the tuples corresponding to the objects from the Ω area, the dataset $W = U \setminus G$ was obtained, the number of tuples in which is less than the number of tuples in the corresponding dataset U.

Table 1. The values of the classifier parameters and the values of the classification quality indicators for datasets (the case of successful application of the proposed classification method).

Dataset $s \times q$	SVM classification AUC_{test}/ Number of errors: Er_{train}/ Er_{test}	Classification type	d_Ω	N_Ω	kNN classification – Method of voting: unweighted – Number of neighbors	unweighted – Number of errors	weighted – Number of neighbors	weighted – Number of errors	with the kernel function – Number of neighbors	with the kernel function – Number of errors	Classification quality ratings: F-measure, %	Accur, %	Se, %	Sp, %	Number of errors: Er_I	Er_{II}	Total
1	2	3	4	5	6	7	8	9	10	11	12	13	14	15	16	17	18
Heart	0.8962/	SVM	–	–	–	–	–	–	–	–	96.67	96.30	96.67	95.83	5	5	10
270 × 13	1 of 216/	+aSim	0.542	35	**37**	**5**	25	6	**37**	**5**	**98.34**	**98.15**	98.67	97.50	2	3	**5**
	9 of 54	+Sim	0.780	43	17	6	31	7	17	6	98.00	97.78	98.00	97.50	3	3	6
Ionosphere	0.9925/	SVM	–	–	–	–	–	–	–	–	97.67	98.29	100	97.33	0	6	6
351 × 34	1 of 281/	+aSim	0.207	9	**11**	**3**	13	3	**11**	**3**	**98.80**	**99.15**	98.41	99.56	2	1	**3**
	5 of 70	+Sim	0.414	16	**11**	**3**	13	3	**11**	**3**	**98.80**	**99.15**	98.41	99.56	2	1	**3**
Portu	0.7211/	SVM	–	–	–	–	–	–	–	–	98.74	97.84	100	86.00	0	14	14
649 × 30	0 of 518/	+aSim	0.671	107	**11**	**5**	12	5	**11**	**5**	**99.00**	**98.30**	100	89.00	0	11	**11**
	14 of 129	+Sim	1.314	107	**11**	**5**	12	5	**11**	**5**	**99.00**	**98.30**	100	89.00	0	11	**11**
Math	0.5719/	SVM	–	–	–	–	–	–	–	–	96.54	95.19	100	85.38	0	19	19
395 × 30	0 of 316/	+aSim	0.300	76	**15**	**13**	16	13	**15**	**13**	**97.24**	**96.20**	99.62	89.23	1	14	**15**
	19 of 79	+Sim	0.600	76	**15**	**13**	16	13	**15**	**13**	**97.24**	**96.20**	99.62	89.23	1	14	**15**
Parkinsons	0.9949/	SVM	–	–	–	–	–	–	–	–	99.32	98.97	99.32	97.92	1	1	2
195 × 23	0 of 156/	+aSim	0.075	2	**0**	**5**	**0**	**5**	**0**	**5**	**100**	**100**	100	100	0	0	**0**
	2 of 39	+Sim	0.105	2	**0**	**5**	**0**	**5**	**0**	**5**	**100**	**100**	100	100	0	0	**0**
LSVT Voice	0.7460/	SVM	–	–	–	–	–	–	–	–	90.91	94.44	83.33	100	7	0	7
126 × 309	0 of 101/	+aSim	0.307	24	3	11	**2**	**13**	3	11	96.47	97.62	97.62	97.62	1	2	3
	7 of 25	+Sim	0.614	24	**1**	**13**	**1**	**13**	**1**	**13**	**98.80**	**99.21**	97.62	100	1	0	**1**
Musk (Version 1)	0.9594/	SVM	–	–	–	–	–	–	–	–	91.62	93.28	84.54	100	32	0	32
476 × 166	0 of 381/	+aSim	0.082	47	**9**	**5**	11	27	**9**	**5**	97.85	98.11	99.03	97.40	2	7	9
	32 of 95	+Sim	0.164	47	9	1	**7**	**5**	10	3	**98.33**	**98.53**	99.52	97.77	1	6	**7**
German	0.7552/	SVM	–	–	–	–	–	–	–	–	96.33	94.70	99.43	83.67	4	49	53
1000 × 24	0 of 800/	+aSim	1.86	219	**39**	**41**	43	45	**39**	**41**	**97.11**	**95.90**	**98.29**	**90.33**	**12**	**29**	**41**
	53 of 200	+Sim	1.796	225	35	47	47	48	35	47	96.72	95.30	99.00	86.67	7	40	47

Further, for objects of the Ω area, the classification results were refined using *k*NN classifier based on tuples of the set $W = U \setminus G$ with a different number of neighbors (varying from 1 to 51) using various voting methods (namely, weighted and unweighted voting) and various methods of assessing the proximity between objects. For each voting method, the optimal number of neighbors was determined (columns 6, 8, 10 of Table 1), for which the number of classification errors is minimal (columns 7, 9, 11 of Table 1).

After applying *k*NN classifier with the number of neighbors, the proximity assessment method and the voting method that ensure the minimum number of classification errors, for different variants (asymmetric and symmetric) of the location of the Ω area relative to the dividing hyperplane, the quality of the final data classification was again evaluated by calculating the values of various indicators. In addition, the new values of quality indicators were compared with the values of the same indicators obtained on the basis of the developed SVM classifier. The cells in Table 1 corresponding to the best *k*NN classifier that provide the minimum number of classification errors are shown in bold.

Table 1 shows that in most cases the maximum refinement of the classification results is achieved when using the asymmetric Ω area, but in a number of experiments the proposed classification method gives the same results when using both the asymmetric Ω area and the symmetric Ω area, and sometimes the maximum refinement of the results classification is achieved using the symmetric Ω area. It is also seen from Table 1 that it is impossible to unambiguously determine the voting method and the proximity assessment method, which would always achieve a better refinement of the classification results when using *k*NN classifier, since in a number of experiments the best solution takes place with an unweighted voting method, and sometimes with a weighted.

Table 2 presents the indicators values of the F-measure, the accuracy indicator ($Accur$), the total number of errors of I-st and II-nd type $(Er_{I}+Er_{II})$ before and after application of *k*NN classifier, which provided the best refinement of the classification results, for the classification of objects that fell into the Ω area, the symbol δ denotes the difference between the corresponding values of SVM classification indicators and the new classifier obtained when implementing the proposed two-stage classification method.

Table 2. Comparison of classifiers.

Dataset	*F-мера, %*			*Accur, %*			*Er*Er_I+Er_{II}		
	SVM	**SVM + *k*NN**	δ	**SVM**	**SVM + *k*NN**	δ	**SVM**	**SVM + *k*NN**	δ
Heart	96.67	98.34	1.67	96.30	98.15	1.85	10	5	–5
Ionosphere	97.67	98.80	1.13	98.29	99.15	0.86	6	3	–3
Portu	98.74	99.00	0.26	97.84	98.30	0.46	14	11	–3
Math	96.54	97.24	0.7	95.19	96.20	1.01	19	15	–4
Parkinsons	99.32	100	0.68	98.97	100	1.03	2	0	–2
LSVT Voice	90.91	98.80	7.89	94.44	99.21	4.77	7	1	–6
Musk (Version 1)	91.62	98.33	6.71	93.28	98.53	5.25	32	7	–25
German	96.33	97.11	0.78	94.70	95.90	1.2	53	41	–12
Average			2.5			2.1			–7.5

Table 3. Values of the classifier parameters and values of the classification quality indicators (case of unsuccessful application of the classification method).

Dataset $s \times q$	SVM classification								Asymmetric Ω area			Symmetric Ω area		
	F-measure, %	*Accur*, %	AUC_{test}	Number of errors Er_{train}	Number of errors Er_{test}	Number of errors *Total*	$Er_{\lvert f \rvert<1}$	$Er_{\%}$	d_{Ω}	N_{Ω}	Number of objects for training the kNN classifier	d_{Ω}	N_{Ω}	Number of objects for training the kNN classifier
Pima Indians Diabetes 768 × 8	83.99	88.28	0.8148	54 of 615	36 of 153	90	82	91	2.612	721	47	2.828	727	41
Liver 345 × 6	81.79	84.64	0.8034	37 of 276	16 of 69	53	47	89	2.837	328	17	2.869	329	16
Spam 4601 × 57	95.24	96.39	0.9490	27 of 3681	139 of 920	166	153	92	1.981	2940	1661	2.373	4512	89
Vertebral 310 × 6	91.13	88.39	0.9810	33 of 248	3 of 62	36	31	86	2.146	208	102	2.17	216	94

Table 3 shows the values of the parameters of the classifiers and the values of the classification quality indicators for the datasets in the case of the unsuccessful application of the proposed classification method, that is, if it is impossible to improve the classification results using *k*NN algorithm. Here $Er_{\lvert f \rvert<1}$ is the number of errors inside the dividing line, $Er_{\%}$ is the share of the *Total* number of errors (in percent). It can be seen from Table 3 that for the datasets presented in it, the Ω area has a significantly wider width than the Ω area for the datasets from Table 2, which was one of the reasons for the unsuccessful application of *k*NN algorithm.

4. Conclusion

Based on the results of the experimental studies, it can be concluded that the use of the proposed method improves the quality of the classification results, since application of kNN classifier to objects located near the hyperplane separating the classes and determined by SVM classifier reduces the number of erroneously classified objects. The proposed classification method allows to make highly accurate decisions on the classification of complex multidimensional data.

References

[1] Hastie T, Tibshirani R and Friedman J 2009 Springer 533 p
[2] Cortez P and Silva A 2008 *Proceedings of 5th FUture BUsiness TEChnology Conference (FUBUTEC 2008)* ed A Brito and J Teixeira 5–12
[3] Chesalin A N, Grodzenskiy S Ya, Nilov M Yu and Agafonov A N 2019 *Russian Technological Journal* (in rus) **7**(5) 20–9 https://doi.org/10.32362/2500-316X-2019-7-5-20-29
[4] Xiaohua Hu, Lin T Y and Louie E 2003 *Proceedings of the 2003 International Database Engineering and Applications Symposium*
[5] Ciresan D C, Meier U, Gambardella L M and Schmidhuber J 2010 *Neural Computation* **22** (12) 3207–20 http://dx.doi:10.1162/neco_a_00052
[6] Hall P, Park B U, Samworth R J 2008 *Annals of Statistics* **36** (5) 2135–52 http://dx.doi:10.1214/07-AOS537
[7] Nigsch F, Bender A, van Buuren B, Tissen J, Nigsch E and Mitchell J B 2006 *Journal of Chemical Information and Modeling* **46** (6) 2412–22 http://dx.doi:10.1021/ci060149f
[8] Wang H and Bell D 2004 *The Computer Journal* **47** (6) 662–72
[9] Vapnik V 1998 *Statistical Learning Theory* (New York: John Wiley & Sons) 732 p
[10] Chapelle O, Vapnik V, Bousquet O and Mukherjee S 2002 *Machine Learning* **46** (1–3) 131–59 http://dx.doi.org/10.1023/A:1012450327387
[11] Bottou L and Lin C-J 2007 MIT Press 1–28
[12] Demidova L and Sokolova Yu 2016 SVM Classifiers: The Objects Identification on the Base of Their Hyperspectral Features *ITM Web of Conferences* **10** 02003
[13] Zhang H, Berg A C, Maire M and Malik J 2006 *Proceedings – 2006 IEEE Computer Society Conference on Computer Vision and Pattern Recognition* **2** 2126–36
[14] Li R, Wang H-N, He H, Cui Y-M and Du Zh-L 2007 *Chinese Journal of Astronomy and Astrophysics* **7** (3) 441–7
[15] Demidova L and Sokolova Yu 2017 *6-th Mediterranean Conference on Embedded Computing* 459–62

IOP Conf. Series: Materials Science and Engineering 1027 (2021) 012002 doi:10.1088/1757-899X/1027/1/012002

Hybrid evolutionary algorithm in the optimization problem of the shooting system parameters

L A Demidova[1] **and N A Petrova**[2]

[1] MIREA – Russian Technological University, Moscow 119571, Russia
[2] Ryazan, Russia

demidova.liliya@gmail.com

Abstract. The optimization problem of the shooting system parameters values by means of the hybrid evolutionary algorithm has been considered. The idea of hybridization of the genetic algorithm and differential evolution algorithm to solve the problem of optimizing the coverage of the observed object by the shooting system has been proposed and implemented. The results of solving the optimization problem of the shooting system parameters, confirming the feasibility application of the proposed hybrid evolutionary algorithm which provides minimization of time spent on solving the optimization problem has been described.

1. Introduction

The solution to the problem of optimizing the coverage of the observation object by the shooting system involves search for the shooting parameters values which provide optimal coverage of this object. The solution to this problem is complicated by the need to carefully select the appropriate criterion and taking into account a large number of restrictions for the values of optimization parameters, as well as restrictions imposed on the optimization criterion. In particular, when solving the problem of optimizing the coverage of an object by the shooting system, such parameters as the latitude of the point of the shooting start, the longitude of the point of the shooting start, the start time of the shooting, the duration of the shooting, the speed of the image movement, the azimuth and restrictions on their values are used. Using the grid search method for the possible values of optimization parameters with subsequent calculation of the value of the optimization criterion in order to select the best set of the required values of the parameters for solving this problem is associated with significant time-consuming, which is not acceptable.

One of the approaches which allows to solve the problem of optimizing the coverage of an object by the shooting system with acceptable time costs assumes the use of the evolutionary optimization algorithms such as genetic algorithm, differential evolution algorithm, particle swarm algorithm, bee colony algorithm, ant algorithm, algorithm clonal selection, etc. In this case, the optimization criterion acts as the fitness function of such algorithms operating in the search space by populations of the possible solutions encoded using optimization parameters. As a result of the iterations implementation of the evolutionary optimization algorithms, a certain locally optimal solution, which is assumed to be the sought one, is determined.

The disadvantage of the evolutionary optimization algorithms is that often the solutions found with their use fall into the local extremum of the fitness functions used, since the initialization of the initial population of solutions is performed randomly.

To derive solutions of some evolutionary optimization algorithm from the local extremum, it is necessary to perform some manipulations with the population of solutions atypical for this algorithm. In particular, the hybridization of various evolutionary algorithms (EA) allows improving the results of solving optimization problems [1–3], thanks to which it is possible to derive solutions from the local extremum and continue the effective development of the population (in the sense of moving towards the global extremum of the optimization criterion (fitness function).

The proposed hybrid evolutionary algorithm (HEA), which intelligently combines the advantages of the genetic algorithm (GA) [1] and the differential evolution algorithm (DEA) [2], which are in many ways similar in the tools used in them, should allow to minimize the risks of "getting stuck" of solutions at the local extremum of the fitness functions.

The problem of optimizing the parameters of the shooting system should be considered as the multi-parameter one-criterion problem of optimizing the values of the parameters which allows to achive the maximum efficiency of the operation of the shooting system.

The solution to this optimization problem is complicated by the fact that for given initial values of the shooting parameters, which should be modified in some way to determine their optimal values, only the permissible ranges of variation of the values of these parameters are known. Also, there is only some algorithm for assessing the quality of the shooting system functioning, and, in addition, the additional requirements and restrictions, which are imposed by the subject area of the optimization problem, are set.

The solution to the problem of optimizing the values of the shooting parameters should be a set of parameter values which meets all the requirements and limitations of this problem and corresponds to the maximum quality of the shooting system, determined by some reasonably chosen optimization criterion.

For the problem of optimization of the shooting parameters formulated in this way, the use of the evolutionary approach to its solution should ensure with the acceptable time costs obtaining the desired solution under a minimum amount of initial data. In this case, it is advisable to perform hybridization of various EA.

2. Theoretical research

The evolutionary approach (EA) to applied optimization problems allows the use of the genetic algorithms, the evolutionary programming, the evolutionary strategies, etc. as the optimization tools, which implement the simultaneous search for alternative solutions and the subsequent selection of the best one at each iteration [4–6].

The use of EA allows to reduce the solution of the optimization problem to the following two generalized steps.

1. Formalization of the optimization problem, as a result of which optimization parameters, ranges of the acceptable values of optimization parameters, optimization criteria (fitness functions) and auxiliary parameters (in particular, constants which are used in calculations) are determined.
2. Optimization and analysis of the results of solving the optimization problem, assuming the launch of EA, determination of the best values of the parameters of the optimization problem for the current iteration of EA and their presentation in the form convenient for analysis.

In recent years, hybridization of various variants of EAs has been actively carried out in order to compensate for their particular shortcomings and obtain more efficient solutions to optimization problems.

When solving the problem of optimizing the parameters of the shooting system, it is proposed to implement the hybridization of GA [1], which is well-known and effectively used in solving various optimization problems, and the DEA [2], less popular, but also found active use in recent years.

Both evolutionary algorithms work with a population of solutions presented in the form of chromosomes, the genes of which encode certain parameters of the optimization problem. In the process of implementing these algorithms, the corresponding populations of solutions are developed using the principles of elitism.

In the basic version of GA algorithm, solutions move in the search space by performing crossover and mutation operations, which modify the coordinates (genes) of solutions from the population.

In GA, the source of noise is an external random number generator which modifies the coordinates (genes) of solutions (chromosomes) of the current population.

In the basic version of DEA, solutions move in the search space according to the law determined by a mathematical formula which provides the modification of coordinates (genes) of solutions from the population. If the newly found solution is better than the existing ones [in the sense of the used optimization quality criterion (fitness function)], it is accepted as a new one and introduced into the population.

In DEA, the source of noise is not an external (as in GA), but an internal random number generator, which is implemented as the difference between the values of genes of randomly selected chromosomes of the current population.

An internal source acts as the noise which distorts the “gene pool” of a chromosome, namely, the difference between the values of coordinates (genes) of the randomly selected chromosomes of population. As a result, a chromosome with original genetic characteristics and a chromosome with genetic characteristics distorted according to the formula are involved in crossing.

The advantage of DEA is its ability to dynamically simulate the relief features of the correspondence function by adapting an internal noise source. This allows DEA to very quickly go out of the extended local minima and ensures the efficiency of its application. At present, various versions of DEA have been proposed which implement the adaptation of the values of its parameters in the optimization process.

It is proposed to use DEA together with GA as the parts of HEA in order to find the best solution to the problem of finding the shooting parameters values with minimal time costs.

In the case of HEA, the population of solutions develops on the basis of several subpopulations of two types of equal size. The subpopulations of one type are formed on the basis of GA, and subpopulations of another type are formed on the basis of DEA. All subpopulations first develop independently of each other for a certain number of generations, and then they exchange the best solutions (with the subsequent removal of the worst solutions to keep the subpopulation size unchanged), after which they develop again independently of each other, etc.

The process of development of subpopulations and exchange of the best solutions ends when a given number of iterations is reached or when the condition for stopping HEA, which is based on the requirement to achieve a predetermined computation accuracy of the value of the fitness function, is met.

When solving the problem of optimizing the coverage of the observation object by the shooting system using HEA, in the general case, it is proposed to use:

- subpopulations of different types, generated randomly;
- a fixed size of subpopulations corresponding to a number of alternative options for solving the optimization problem, which are the sets of the shooting parameters values;
- a fixed number of genes in a chromosome, corresponding to the number of optimized parameters.

It is expedient to take the number of subpopulations of one type equal to the number of subpopulations of another type. Each new subpopulation of chromosomes of size N / K (N is the number of chromosomes in the population, K is the number of subpopulations, which is even) of the current iteration is formed from an auxiliary subpopulation of chromosomes containing parent and offspring chromosomes by removing chromosomes with the worst fitness function values so as to maintain a fixed size of the subpopulation and ensure the replacement of chromosomes in the subpopulation with the "best" ones, the values of genes of which are closer to the optimal values of the parameters of the optimization problem.

Periodically (for example, after a set number of iterations), the best chromosomes are exchanged between subpopulations of different types. When the number of subpopulations K>2, the subpopulations participating in the exchange of the best solutions are chosen randomly from the number of subpopulations of different types.

In addition, the exchange of the best solutions can be carried out in a loop, in which, for example, subpopulations of different types are ordered through one.

This approach to the exchange of the best chromosomes makes it possible to implement a kind of parallel EA, which has some properties of the island model, suggesting the existence of a group of the "closely spaced islands" with evolving subpopulations of the same type, developing independently and only occasionally exchanging chromosomes.

The considered island model assumes the existence of several subpopulations of the same fixed size; a fixed number of genes on a chromosome; the use of arbitrary variants of strategies for the selection and formation of generations in subpopulations; no restrictions on crossing and mutation operations; exchange of chromosomes between subpopulations.

3. Experimental results

To check the operability of HEA and the optimization software system based on it, the applied problem of optimizing the coverage of the observation object by the shooting system was considered [7–9].

Figure 1 shows the initial boundaries of the shooting, calculated from the initial values of the parameters (the dark rectangle is the object; the light polygon is the capture of the shooting system).

As a criterion for optimal coverage of the observation object, it is advisable to use the minimum value of the difference between the coverage area of the shooting and the area of the object.

To solve the problem of optimizing the parameters values of the shooting system, the above-described HEA was applied.

Previously, the subject area of the optimization problem was presented in terms of EA.

The solution to the problem of optimizing the coverage of the observation object by the shooting system was presented in the form of the chromosome, in which the parameters of the problem, which are coordinates in the search space, are encoded.

If, for example, the number of subpopulations is 2, that is, K = 2, then the i-th chromosome in the subpopulation of solutions of size $N/2$ can be written as [7]:

$$P_i = \left(nshir_i, ndolg_i, ontel_i, dlit_i, sdi_i, azim_i\right), \tag{1}$$

where P_i is the i-th chromosome $(i = \overline{1, N/2})$; $nshir_i$ is the latitude of the shooting start point; $ndolg_i$ is the longitude of the shooting start point; $ontel_i$ is the shooting start time; $dlit_i$ is the duration of shooting; sdi_i is the image movement speed; $azim_i$ is the azimuth; $N/2$ is the subpopulation size; N is the population size.

For each gene in the chromosome P_i $(i = \overline{1, N/2})$, it is advisable to determine (in accordance with the restrictions imposed by the subject area of the optimization problem) the range of the acceptable values (Table 1).

During the experimental research, the planning of the shooting of the observation objects was carried out and the initial values of the parameters of the shooting system for the selected object were determined. Table 1 shows the initial values of the optimization parameters, the intervals of the acceptable values of the optimization parameters for the selected object.

As the genes of the chromosome in the applied problem under consideration, it is proposed to use the latitude of the shooting start point, the longitude of the shooting start point, the shooting start time, the shooting duration, the image movement speed and the azimuth.

IOP Conf. Series: Materials Science and Engineering 1027 (2021) 012002 doi:10.1088/1757-899X/1027/1/012002

Figure 1. The shooting boundaries which are calculated from the initial parameters values.

Table 1. The initial values of the shooting parameters, ranges of the acceptable values of the shooting parameters and the names of genes in the i-th chromosome.

Shooting parameter	Initial value	Range of the acceptable values
Latitude of the shooting start point, deg. ($nshir_i$)	49.7843	(40; 50)
Longitude of the shooting start point, deg. ($ndolg_i$)	52.4260	(47; 58)
Shooting start time, sec. ($ontel_i$)	2024411	(2024350; 2024470)
Shooting duration, sec. ($dlit_i$)	7.90497	(5; 15)
Image movement speed, meters per second (sdi_i)	60	(55; 65)
Azimuth, deg. ($azim_i$)	0	(0; 360)

Table 2. The generalized sequence of steps of GA and DEA.

Step	GA	DEA
1	To create the initial subpopulation of size $N/2$ from chromosomes P_i ($i=\overline{1,N/2}$).	To create the initial subpopulation of size $N/2$ from chromosomes P_i ($i=\overline{1,N/2}$).
2	As long as the current number g of iterations is less than the maximum number G of iterations, to select the parent chromosomes from the population are randomly and to go to step 3; when the number of iterations is reached, to go to step 5.	As long as the current number g of iterations is less than the maximum number G of iterations, to select the parent chromosomes from the population are randomly and to go to step 3; when the number of iterations is reached, to go to step 5.
3	To create the auxiliary population from parent chromosomes and offspring chromosomes by means of the single-point crossing and (or) mutation, to calculate of the values of the fitness function (2) and to check the constraint (3).	To create the auxiliary population by crossing the direct parent chromosome and the parent chromosome distorted by the special mutation operator, to calculate the values of the fitness function (2) and to check the constraint (3).
4	To create the new subpopulation of size $N/2$ by excluding the chromosomes with the worst fitness function values; to go to step 2.	To create the new subpopulation of size $N/2$ by excluding the chromosomes with the worst fitness function values from the number of the offspring chromosome, the direct parent chromosomes and the distorted parent chromosome; to go to step 2.
5	To select the chromosome with the minimum value of the fitness function (2) from the subpopulation.	To select the chromosome with the minimum value of the fitness function (2) from the subpopulation.

The considered applied optimization problem can be formulated as the problem of finding the optimal values of the shooting parameters: $nshir_{opt}$, $ndolg_{opt}$, $ontel_{opt}$, $dlit_{opt}$, sdi_{opt} , $azim_{opt}$.

Since it is required to ensure the maximum approximation of the shooting area to the area of the observation object so that the vertices of the observation object lie within the boundaries of the shooting, and the shooting area differs from the area of the observation object by a minimum value, the formula for the fitness function of EA can be written as:

$$\left(S_s - S_{ob}\right) + N_{ob} \cdot N_{kr} \to \min, \tag{2}$$

where S_s is the shooting area; S_{ob} is the area of the object; N_{ob} is the number of object vertices which do not fall within the boundaries of the shooting; N_{kr} is the number which is significantly greater than the difference in areas and is necessary to ensure the significant deterioration in the value of the fitness function if the condition that all the vertices of the object fall into the shooting area is not met.

№	nshir	ndolg	ontel	dlit	sdi	azim	criteriy
219	49.549623	52.352618	2024441.666004	9.107331	63.440073	9.540515	**6434.1531510000**
244	49.527458	52.297285	2024442.898216	8.645524	63.660712	14.021995	**5784.0959620000**
256	49.498445	52.296438	2024442.287044	8.602764	63.770609	12.520767	**5719.8944460000**
			....	....	....		
912	49.634652	52.493574	2024424.728382	6.895812	63.608610	2.937738	**2680.7302230000**
1000	49.661975	52.497074	2024423.480220	6.956461	63.513743	2.955700	**2674.5679390000**

Figure 2. The fragments of the dialog box of the optimization program with the sets of the optimized parameters values and the values of the optimization criterion (named as criteriy).

Figure 3. The shooting boundaries calculated according to the optimal parameters values.

The fitness function can be used as the optimization criterion in the problem of optimizing the parameters of the shooting system.

Since the difference between the areas of the object and the shooting will tend to the minimum value of the square kilometers, it is advisable to set the value for the N_{kr} equal to 100000.

The choice of such value for the N_{kr} will provide the increase in the area difference by a factor of hundreds and will make it possible to reject solutions to an applied problem with uncovered object vertices.

Since the shooting area must cover the area of the object, it is advisable to ensure that the positive condition is satisfied for the fitness function:

$$((S_s - S_{ob}) + N_{ob} \cdot N_{kr}) > 0. \tag{3}$$

Table 2 shows the generalized steps of GA and DEA.

If the current number g of iterations is not a multiple of some number I ($I < G$, where G is the maximum number of iterations), then the operations of crossing and mutation are implemented in each subpopulation independently of each other. If the current number g of iterations is a multiple of the number g, then the best chromosomes are exchanged between subpopulations of GA and DEA (when the number of subpopulations K is greater than 2, the exchange of chromosomes occurs in a ring). From subpopulations of GA and DEA algorithms, one chromosome, which has the best fitness function value in its subpopulation, is extracted. To the extracted chromosomes, the operation of the exchange of chromosomes-clones is applied [while in their subpopulations the best chromosomes are preserved, and the worst ones are replaced by chromosomes-clones (if they are really the best)]. A herewith, I is the number of iterations, which makes it possible to increase the efficiency of HEA by means of the redistribution of genes in chromosomes during the exchange of chromosomes between the different subpopulations at each iteration.

Using the proposed HEA, the optimal values of the shooting parameters of the coverage of the observation object were calculated for the given initial values of the parameters and constraints (Table 1). A herewith, the maximum number G of iterations is equal to 1000, the number I of iterations through which the interaction of subpopulations is carried out is equal to 10, the error in reaching the minimum of the fitness function is equal to 0.0001, the number of subpopulations K is equal to 2, the number of offsprings in the crossing operators of DEA is equal to 1, the number of offsprings in the crossing operators of GA is equal to 2, the size of population N is equal to 20, the size of the subpopulations is equal to 10, the mutation strength M in DEA is equal to 0.8, the probability c_{GA} of crossing in GA is equal to 0.6, the probability m_{GA} of mutation in GA is equal to 0.7.

As a result of the operation of HEA on the computer with the Intel Core i7 processor with the clock frequency of 3.07GHz and 1.99GB RAM for 3 minutes, the fitness function (2) was sequentially minimized.

Figure 2 shows fragments of the dialog box of the optimization program based on the proposed HEA, containing the sets of the optimized parameters values and values of the optimization criterion determined on the basis of the fitness function (2).

Figure 3 schematically shows the boundaries of the shooting, calculated according to the obtained parameter values (the dark rectangle is the object; the light polygon is the capture of the shooting system).

The calculation results confirm the advisability of using HEA when solving the problem of optimizing the coverage of the observation object by the shooting system, which assumes that the discrepancy between the object area and the shooting area should be minimal, and the shooting area should cover the object area at the given values of the number of iterations and the accuracy of HEA.

The considered example of solving the problem of optimizing the coverage of the observation object by the shooting system clearly demonstrates the capabilities of the proposed HEA and the developed software product when solving an applied optimization problem.

During the experimental studies, the effectiveness of the application of the program for optimizing the values of the parameters of the shooting system was shown. Using of program allowed to reduce the initial value of the optimization criterion (2) by more than 75.7 times: 2674.567939 at the one thousand iteration against to 202627.913184 at the first iteration.

Earlier, to solve the considered optimization problem, a version of the parallel GA [7] was used, hybridization of which with DEA allows in some cases to reduce (depending on the topological features of the observation object) the search time for optimal values of the shooting system parameters by 5–10% in the context of minimizing the optimality criterion (2): the similar values of the optimality criterion (2) can be obtained with less time.

4. Conclusion

The obtained results of the practical application of HEA confirm its effectiveness in solving the problem of optimizing the parameters values of the shooting system, since it allows to obtain the desired solution with minimal time costs.

In the future, the solution of the problem of the multi-objective optimization of the parameters values of the shooting system by means of EA [10] is of considerable interest.

In addition, it is planned to consider the solution of the problem of optimizing the parameters values of the shooting system, if it is necessary to cover a group of objects by shooting.

References

[1] Anfyorov M A 2019 *Rossiiskii tekhnologicheskii zhurnal = Russian Technological Journal* **7**(6) 134–150 (In Russian). https://doi.org/10.32362/2500-316X-2019-7-6-134-150
[2] Chakraborty U K 2008 *Advances in Differential Evolution*
[3] Demidova L A and Gorchakov A V 2020 *Symmetry* **12**(5) 784 10.3390/sym12050784.
[4] Krus P K and Ölvander (Andersson) J 2013 *Engineering Optimization* **45**(10) 1167–1185
[5] Afonin P V 2009 *Proceedings of the XV Int. Conf. "Knowledge-Dialogue-Solution"* **1** 81–84
[6] Younis A and Dong Z 2010 *Engineering Optimization* **42**(8) 691–718
[7] Demidova L A, Petrova N A and Nikulchev E V 2017 *Contemporary Engineering Sciences* **10**(6) 253–262
[8] Demidova L A, Petrova N A and Sablina V A 2018 *7th Mediterranean Conf. on Embedded Computing (MECO)* 182–185
[9] Demidova L A, Petrova N A and Gazizova O R 2017 *IOP Conf. Series: Materials Science and Engineering* **312** 012004
[10] Deb K, Pratap A, Agarwal S and Meyarivan T 2002 *IEEE Transactions on Evolutionary Computation* **2** 182–197

IOP Conf. Series: Materials Science and Engineering 1027 (2021) 012003 doi:10.1088/1757-899X/1027/1/012003

Synergy of QFD and FMEA methods to improve workplaces in aircraft wire manufacturing

A Ya Dmitriev[1], **T A Mitroshkina**[1] **and I V Russkikh**[1,2]

[1] Samara National Research University named after academician S.P. Korolev (Samara University), 34 Moskovskoye Shosse, Samara 443086, Russia
[2] JSC "Samara Cable Company" (JSC "SCC"), 9 Kabelnaya ul., Samara 443022, Russia

dmitriev57@rambler.ru

Abstract. The integration of QFD and FMEA methods for improving workplaces is considered. The use of priority calculations in the QFD method in matrix form facilitates integration with the FMEA risk analysis method. An example is given of improving the work places of supporting operations in the manufacture of dual-use wire & cable products. It is proved that the direction of improving the activities and workplaces in the enterprise management system, determined on the basis of the proposed synergistic FMEA and QFD methods, is an innovative way of developing a wire&cable manufacturing.

1. Introduction

Aviation airborne wire is a product that has special requirements due to the specifics of operation. Designed for power and control circuits of instruments and electrical equipment installed on board an aircraft, such a wire must combine high reliability and the lowest possible weight and dimensions, as well as have properties that exclude the possibility of damage during installation and during operation (increased resistance to bending, abrasion and bursting). One of the innovative and promising developments in the aerospace and automotive industries is the Spec55 series wires. They have proven themselves well in many aviation programs around the world, from small businesses to such global giants as Boeing and Airbus.

The importance of using modern quality management tools, including the concept of "Lean Manufacturing" in the development of innovative products and technologies, is determined by the long-term practice of the best enterprises and international management standards in the automotive and aviation industries. However, in Russia, such tools as quality function deployment (QFD) and failure mode and effects analysis (FMEA) have so far found a superficial application and their role in the development of innovations and implementation of lean production is underestimated [1–3]. There has been unacceptably little emphasis on deploying and applying these tools to deliver the right quality right in the workplace.

In the generally accepted sense, innovation is an implemented innovation in the field of technology, technology, labor organization or management, based on the use of scientific achievements and best practices, providing a qualitative increase in the efficiency of a system or the quality of products and services. The structure of innovations of Russian enterprises practically does not change from year to year — the greatest attention is paid to the acquisition of machinery and equipment (Figure 1).

Small improvements (kaizen) can be developed both on a planned basis and as a result of FMEA in the form of measures such as finalizing the equipment, introducing error protection elements (poka-yoke), introducing the "Sequencing" system (5S), etc. Innovations, including radical changes in the management and infrastructure of the organization (the introduction of technologies, automated control systems and world-class equipment), can be developed as a result of the implementation of the QFD method.

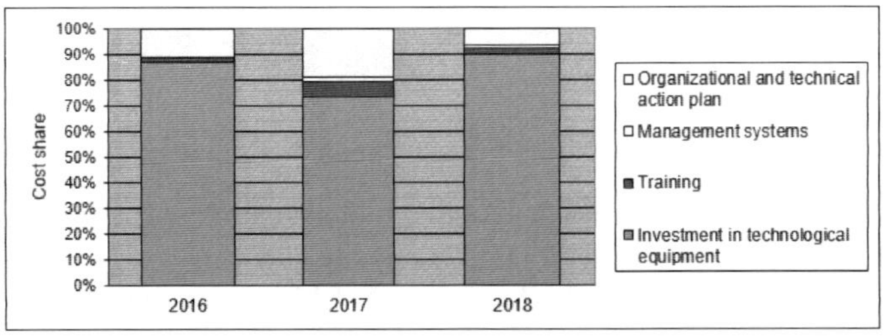

Figure 1. Innovative directions at JSC "SCC".

A methodology for the systematic and structured transformation of customer requirements into product requirements and the further deployment of requirements down to the workplace with the least risk is a modern instrumental way of developing innovations.

2. Examples of the implementation of modern approaches to quality assurance in the aerospace industry

2.1. Structuring and deploying customer requirements

The quality function deployment (QFD) method, along with planning experiments and other methods, is an element of the modern approach to quality management "Big Q" [1, 7]. Despite the fact that the QFD method has long been used by various companies in Japan and the USA, and is widely implemented in Europe, it can be noted that approximately 80% of publications on Russian devoted to its implementation contain descriptions of only Level I QFD.

The QFD method is implemented using a matrix diagram, the so-called House of Quality (HOQ). In its expanded form, QFD includes four phases, and for each of them its own HOQ is built. After converting consumer characteristics into technical ones (phase No. 1, QFD level I), the latter are converted into characteristics of components (phase No. 2, QFD level II), then into process parameters (phase No. 3, QFD level III), and then into requirements to the execution of operations (phase 4, QFD IV level).

The absolute values of the characteristics' priorities in matrix form are determined by the formula:

$$\Theta = H^T \times P, \tag{1}$$

where H^T is the transposed matrix of relationships (for QFD I level — the transposed matrix of relationships between consumer and technical characteristics); P — vector of weights (for QFD I level — vector of weights of consumer characteristics).

2.2. Risk analysis method

Risk analysis using the failure mode and effects analysis (FMEA) method is carried out to analyze potential inconsistencies, their causes and consequences, assess the risks of their occurrence and take measures to eliminate or reduce the likelihood of damage from their occurrence [5, 6]. This is one of the most effective methods for finalizing the design of technical objects and their manufacturing processes at such critical stages of the product life cycle as its development and preparation for production.

In the automotive industry, the most common FMEA for design, process and equipment. In this case, the risk is assessed according to the value of the Risk Priority Number (RPN), which is a generalized quantitative characteristic of the nonconformity, taking into account its significance and the likelihood of occurrence and detection of the cause of the nonconformity. A significant drawback of the FMEA process in practical application is that often the production and support operations are not clearly separated or the support operations FMEA is not carried out, since they are often considered not adding value and not requiring attention.

2.3. Integration of QFD and FMEA methods

When implementing the QFD method in the development of new or modernization of existing products, it is proposed to use the results of FMEA analyzes of similar products and processes, expertise of experience and knowledge.

Thus, when the QFD and FMEA methods are used together, a synergistic effect is achieved and the development of innovative products and technological processes is carried out taking into account the risks. The absolute value of the priority of the characteristic using the results of the FMEA methods in the QFD I level (transformation of requirements and expectations of consumers into technical characteristics of products) is calculated in matrix form, taking into account the risk assessment:

$$\Theta = H^T (P \times R^T), \quad (2)$$

where H^T is the transposed matrix of the relationship between characteristics and requirements; P — vector of requirements weights; *R* is the vector of risk assessments.

This approach is extended to the rest of the QFD phases (levels II, III, and IV). This paper presents the results of the implementation of QFD and FMEA methods at level IV (transformation of process parameters into requirements for the execution of operations) - in order to ensure quality directly at the workplace, taking into account measures for the introduction of lean manufacturing technologies.

3. Application of integration of QFD and FMEA methods to improve the efficiency of workspaces of supporting operations in Aircraft Wire Manufacturing

The workplace can be considered as a sub-process of the organization's production process with the application of management functions to it and, therefore, the implementation of the process approach is the basis for the implementation of methods for improving the workplace, in particular, lean manufacturing methods. The workplace is a link in production, the organization and equipment of which determines the efficiency of the use of material and labor resources. Opportunities to reduce waste in the workplace can be found in avoiding overuse of resources. Proceeding from this, the reserves of increasing the efficiency of the use of jobs should be understood as the possibility of using jobs with the use of more economical methods, which, in fact, are the methods of lean production.

In the modern sense in the framework of the process approach, the term "workplace" means: an organizationally indivisible element of an enterprise, in which the forces of one or more employees perform the specified operations of the business process in accordance with the requirements of the consumer. Workplace activities can be viewed as a process model with the allocation of three groups of processes: production, support and management. Manufacturing processes form the foundation for the formation of product value and determine the outputs of the organization. These processes are strategically important to the organization's successful business and directly affect customer satisfaction. In contrast to manufacturing, supporting processes and management processes only indirectly add value to products.

The standard lean approach is to reduce the time for operations that do not add value to the customer. We have analyzed the workplaces of the process of "Production of automotive cable" and "Production of aerospace wire&cable" from the point of view of actions that add and do not add value for the consumer. As a result, it was revealed that only individual actions create value that is important for the consumer. A number of other actions are also needed, although they do not create value. For example, a crimper must change the output and take-up drums and clean the forming head.

From the point of view of the new approach, the task is to spend as little time as possible on actions that do not add value to the product, but without losing the accuracy of meeting requirements and without risk. This is achieved by integrating the QFD and FMEA methods for the workplace.

The most critical stage that determines the quality and performance of the wire is the process of applying insulation to the copper conductor - extrusion. Insulation is carried out on a complex machine called an extrusion line. The most important part of the extrusion line is the die. It is designed to apply insulation consisting of the following elements: inner layer, outer layer, one or more strips of auxiliary color, or surface painting of the outer layer of insulation. A sketch of the extrusion unit is shown in Figure 2a. The appearance of the main element (mandrel) is shown in Figure 2b.

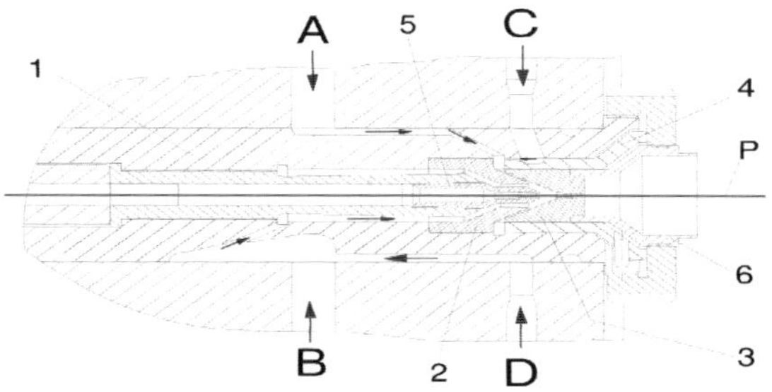

(a)

1 — splitter shaft, 2 — intermediate die, 3 — main die, 4 — strip matrix, 5 — mandrel, 6 — die screw

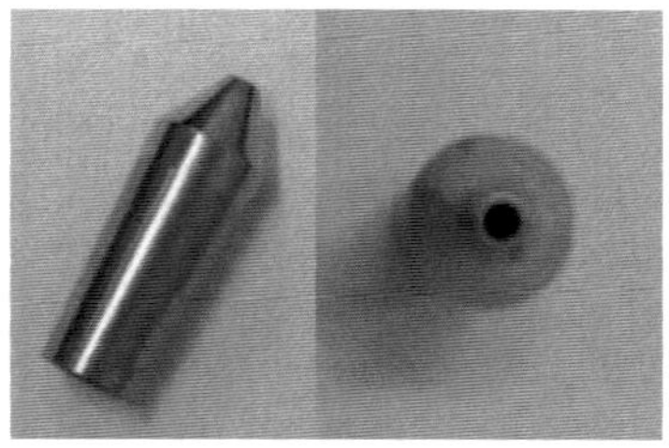

(b)

Figure 2. Extrusion head: (a) sketch of the cross-section of the extrusion head; (b) mandrel.

During operation, the mass from the main extruder A flows along the splitter shaft (pos. 1) between the intermediate matrix (pos. 2) and the matrix (pos. 3), forming the outer layer (the main color of the insulation). The mass from the auxiliary extruder C is supplied to the main mass flow through the channels of the strip die (pos. 4) between the intermediate die and the die (pos. 2 and 3). The inner layer consists of the mass of extruders B and D. These two mass flows are combined in the channels of the splitter shaft (pos. 1) and are directed to the core between the mandrel (pos. 5) and the intermediate die (pos. 2). The inner layer has a mixed color B and D. The die screw (pos. 6) is used to adjust the die position (pos. 3) — the gap between the die and the intermediate die. The feed channels of the mass from a pair of extruders A, C and B, D run symmetrically along the splitter shaft, therefore the insulation layers (inner and outer) are changed by turning the camshaft by 180°. During the color change, pressure peaks are observed in the die. These peaks are absorbed by two discharge chutes on the splitter shaft, from where material exits the head through two discharge holes.

The mandrel hole diameter (Figure 2) is a geometric characteristic that affects the quality of the insulation application, ensures its concentricity. The diameter of the mandrel hole is measured using a Large Instrumental Microscope (BMI-1). The resolution of the measuring instrument is 0.005 mm.

A risk analysis was carried out using the FMEA method for both production and support operations carried out at the workplace of the pressure operator. The values of the RPN have been determined for each type of potential nonconformity. In accordance with the requirements of consumers with a RPN > 100, corrective actions have been developed to reduce this indicator.

Subsequently, when conducting a Level IV QFD for a pressure operator's workplace, the participation of adjusters, mechanics, power engineers and electronics in ensuring the effective functioning of the production process, as well as the results of FMEA of supporting operations, was also taken into account. The results of QFD (Figure 3) demonstrate that the greatest influence on the parameters of the technological process of applying insulation to a conductive conductor is exerted by: no temperature difference between mandrel and mandrel when cleaning the extruder head by an operator.

Row #	Max Relationship Value in Row	Relative Weight	Weight / Importance	Quality Characteristics (a.k.a. "Functional Requirements" or "Hows") / Demanded Quality (a.k.a. "Customer Requirements" or "Whats")	Operator. Head cleaning. Installation temperature of the camshaft in the cylinder liner	Operator. Head cleaning. Head installation and alignment temperature	Operator. Head cleaning. Temperature difference between mandrel and mandrel holder	Service technician. Mandrel diameter control. Diameter (microscope)	Service technician. Mandrel diameter control. Control frequency	Mechanic. Checking the condition of the belts and their tension. Tension	Mechanic. Checking the condition of the belts and their tension. Check frequency	Mechanic. Checking the oil level in the extrusion unit. Oil level deviation	Mechanic. Checking the oil level in the extrusion unit. Check frequency	Energetic. Checking the thermal conditions of the engines. Temperature	Energetic. Checking the thermal conditions of the engines. Check frequency	Electronics. Cleaning controller contacts (power supply unit, CPU, input-output modules). Cleaning frequency
1	1	18,3	146,0	Matrix diameter when applying insulation to a conductive core		▲										
2	9	42,8	342,0	Mandrel diameter	▲	▲	Θ	Θ	O							
3	9	14,1	113,0	Color dosing system								O	▲			Θ
4	9	12,6	101,0	Extrusion line tension						Θ	O			Θ	O	
5	3	12,3	98,0	Temperature conditions for processing materials on the extrusion line								O	▲			
				Target or Limit Value	cold shaft into a heated sleeve (100C)	Working temperature	0	1.17mm	1 per day	50 N (wire tension)	1 time per month	0	1 time per month	Working temperature	Once every 6 months	Once a year
				Difficulty (0=Easy to Accomplish, 10=Extremely Difficult)	1	9	9	6	3	9	10	10	10	1	6	10
				Max Relationship Value in Column	1	1	9	9	3	9	3	3	1	9	3	9
				Weight / Importance	42,8	61,0	384,8	384,8	128,3	113,6	37,9	79,1	26,4	113,6	37,9	127,1
				Relative Weight	2,8	4,0	25,0	25,0	8,3	7,4	2,5	5,1	1,7	7,4	2,5	8,3

Figure 3. Application of the integrated method for the workstation of the pressure operator.

A study of the operating instructions of the operator and the adjuster was carried out, taking into account the risks, importance and influence of these supporting actions on the parameters of the technological process and, ultimately, on the quality of the finished product. Innovative transformations, including the revision of operations and the development of modern illustrated work instructions, allowed CJSC "SCC" to significantly reduce the level of external defectiveness and polyethylene waste (Figure 4).

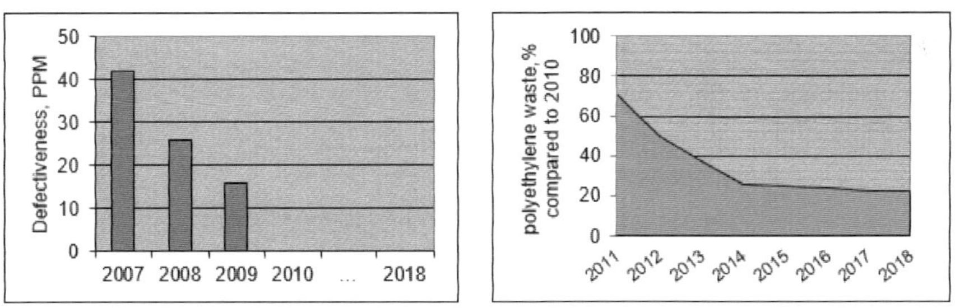

(a) (b)

Figure 4. Dynamics of improvements: (a) external defectiveness, ppm, (b) reduction of polyethylene waste.

As a result of improving workplaces, taking into account the risks identified for the supporting operations, the so-called "non-value-adding" in lean manufacturing, we received a reduction in the cost of performing actions that do not add value to the consumer:

- workplace of a pressure operator for wire — 17.4%.
- workstation of the operator: combined process of drawing copper wire and applying insulation — 18.7%;
- workstation of the twist: twisting of 5 (10) paired bundles — 8.6%.

The presented model is based on a new approach to the design and manufacture of products — bionic design and on approaches to long-term planning of product quality adopted in the aerospace industry. The model is aimed at reducing costs and non-production losses both at the early stages of design and at the stages of production and testing, based on modern QFD and FMEA quality assurance methods and additive production approaches.

4. Conclusions and discussion

The use of QFD and FMEA quality management methods in the development of innovative wire&cable products has been substantiated. Structuring the requirements and expectations of consumers to determine the key innovative characteristics of products, as well as the parameters of technological processes to determine the key innovative technologies based on these methods, increase the efficiency of choosing direction for improving workplaces.

It is proved that the direction of improving the activities and workplaces in the enterprise management system, determined on the basis of the proposed synergistic FMEA and QFD method, is an innovative way of developing a wire&cable manufacturing. This gives the enterprise great opportunities in a competitive environment.

References

[1] Rodionov V N, Popova T V, Dmitriev A Ya, Mitroshkina T A Innovative transformation of workplaces based on QFD and FMEA integration methods *Quality management methods* 2011 8 p 30–35

[2] Rodionov V N, Popova T V, Dmitriev A Ya, Mitroshkina T A Method of development of innovations taking into account risks in the production of auto-tractor wires *Cables and wires* 2011 1 p 10–14

[3] Dmitriev A Ya, Zagidulin R S and Mitroshkina T A 2020 Special Aspects of Quality Assurance in the Design, Manufacture, Testing of Aerospace Engineering Products *IOP Conf. Ser.: Mater. Sci. Eng.* **714** 012006 https://doi.org/10.1088/1757-899X/714/1/012006

[4] Akao Y and Mazur G H 2003 The leading edge in QFD: past, present and future *International Journal of Quality & Reliability Management* 20 1 p 20–35 https://doi.org/10.1108/02656710310453791

[5] *Potential failure mode and effects analysis (FMEA) Reference Manual* 2008 Fourth Edition (Chrysler LLC, Ford Motor Company, General Motors Corporation) 150 p

[6] *ARP5580 Recommended Failure Modes and Effects Analysis (FMEA) Practices for Non-Automobile Applications* 2001

[7] Bhote Keki R and Bhote Adi K 2000 *World class quality: using design of experiments to make it happen* (NY: AMACOM) 487 p

IOP Conf. Series: Materials Science and Engineering 1027 (2021) 012004 doi:10.1088/1757-899X/1027/1/012004

The predictive display as a means of suppressing the negative effects of wake vortex encounters

A V Efremov, M S Tiaglik and T V Voronka

Moscow aviation institute (National Research University), 4 Volokolamskoe Shosse, Moscow 125993, Russia

pvl@mai.ru

Abstract. The research develops a predictive information display system as a means of suppressing the negative effects of wake vortex encounters by aircraft during the landing phase. A mathematical model of aircraft interaction with a wake vortex is developed, with the effectiveness of the predictive information indication being evaluated via studies using a flight simulator.

1. Introduction

The most important requirement considered when designing aviation technology is increasing flight safety. Safety is one of the primary challenges formulated by the International Council of the Aeronautical Sciences (ICAS), the United States Federal Aviation Administration (FAA), as well as the Russian aviation technology development program. All of these documents declare the necessity of reducing the number of human factors-related accidents by 4 times.

Analysis of aviation accident statistics for the past 30 years shows that crashes caused by human factors comprise 80–85%. Of these, 65–75% are related to flight crew errors. According to the statistics given in [1], the primary factors of pilot errors are:

- errors related to manual control;
- omission of actions or incorrect actions;
- disorientation.

In general, flight accidents do not arise from a single factor, but are caused by several factors. Such factors include control system failures, suddenly changing the controlled element dynamics, as well as atmospheric disturbances: turbulence, continuous strong winds, windshear, and vortices, including those caused by previously passing aircraft.

Encountering vortices rarely leads to emergency situations in heavy aircraft, whereas flight incidents followed by passenger or crew injuries in light aircraft are a rather common occurrence. Encountering a vortex or windshear significantly complicates the piloting process, as it leads to a sudden and substantial change in attitude, airspeed and altitude, likelihood of entering a critical flight mode or a stall. In these circumstances, the pilot must quickly perform actions in multiple control channels, which significantly increases his workload.

2. The model of aircraft interaction with a wake vortex

A turbulent wake vortex caused by a previously passing aircraft is a combination of vortices formed by wake flows from various aircraft surfaces.

Aside from heavy workload, the need to achieve precision piloting induces stress, accompanied by a decrease in control precision, and accounts for a reduction in flight safety.

In order to minimize the negative effects of wake vortices, equipped airfields have controllers warning pilots about potential encounters. In this case the pilot follows the glidepath with a speed 5 knots higher and increase the glidepath angle to 3 degrees.

In this study, a four-vortex wake model, developed in [2], was used. Assuming that the vortex is constant over a period of 12–160 seconds and for providing the determinateness of the vortex action on the aircraft (providing the stationarity of external disturbances during the experiment), such a model allows to calculate three projections of wind speed in an arbitrary point (x, y, z) for the position of the vortex, described by the coordinates $y_0 = f(x)$ and $z_0 = f(x)$.

When modeling a wind caused by a vortex, it is important to take into account that in different points in space and, consequently, for different parts of the aircraft lifting surface, the wind speed is different. In this connection, the study suggests dividing the wing and tailplane of the aircraft into segments and calculating the speed and direction of wind gust and associated aerodynamic forces for each of their attitudes.

In accordance with the proposed method, the wing and tailplane were divided into 500 segments for calculating the forces and moments occurring in these segments, induced by wind action. The coordinates of these segments were calculated in the body-fixed axis system, taking into account the balance of the considered aircraft.

Since the mathematical model of wind disturbances calculates the wind speed in an arbitrary point in the normal earth axis system, determining the wind speed in each segment of the wing lifting surface requires transferring the segments coordinates into the normal earth axis system in accordance with the coordinates and attitude of the aircraft. Then, wind components are calculated for each segment position in the normal axis system.

The calculated wind speed projections for each segment of the wing and tailplane are projected into the body-fixed axis system. Knowing the proper motion speed of each wing segment and the wind speed, the angles of attack and sideslip α_T and β_T were determined, induced by wind action in each of the segments.

Based on the initial angles of attack and sideslip α and β, the angles induced by the wind α_T and β_T, as well as on the aerodynamic characteristics, the increments of aerodynamic forces were determined, in accordance with the relations:

$$\Delta F_{xc} = qS_c(C_x(\alpha_T) - C_x(\alpha)),$$

$$\Delta F_{yc} = qS_c(C_y(\alpha_T) - C_y(\alpha)),$$

$$\Delta F_{zc} = qS_c(C_z(\alpha_T) - C_z(\alpha)).$$

After calculating the forces increments in each segment, the increments of moments were calculated.

$$\Delta M_{xc\,i} = X \cdot \Delta F_{z\,i} - Z \cdot \Delta F_{y\,i},$$

$$\Delta M_{yc\,i} = X \cdot \Delta F_{x\,i} - Z \cdot \Delta F_{z\,i},$$

$$\Delta M_{zc\,i} = X \cdot \Delta F_{y\,i} - Z \cdot \Delta F_{x\,i}.$$

Here, X, Y, Z are the coordinates of the wing and tailplane segments in the body-fixed axis system.

In order to use the obtained increments in the mathematical model of aircraft motion, the calculated increments of forces and moments are added up.

$$\Delta F_x = \sum_{i=1}^{500} \Delta F_{x\,i},\ \Delta F_y = \sum_{i=1}^{500} \Delta F_{y\,i},\ \Delta F_z = \sum_{i=1}^{500} \Delta F_{z\,i};$$

$$\Delta M_x = \sum_{i=1}^{500} \Delta M_{xc\,i},\ \Delta M_y = \sum_{i=1}^{500} \Delta M_{yc\,i},\ \Delta M_z = \sum_{i=1}^{500} \Delta M_{xc\,i}.$$

The developed mathematical model of interaction between the aircraft and atmospheric disturbances, unlike the standard models where wind disturbances act on the center of pressure, allows to take into account the different wind action on different parts of the aircraft lifting surface, both in terms of force and direction.

3. Developing the means of suppressing the negative effects of wake vortex encounters

This study considers the predictive display as one such means, with a programmed path motion (a glidepath with the glidepath exit segment up to touchdown) presented on the screen as a three-dimensional corridor with a predictive window moving in the corridor at the speed of the aircraft and located at a distance (or predictive time T_{np}) chosen so that it ensures the minimum variance of glidepath tracking error.

In accordance with the method described in [3], the selection of the time T_{np} was carried out considering the compensatory pilot-aircraft system. The time is determined by the distance $L_{np} = T_{np}V$ between the pilot's eye and a plane (predictive window) located in the corridor covering the programmed path motion (Figure 1).

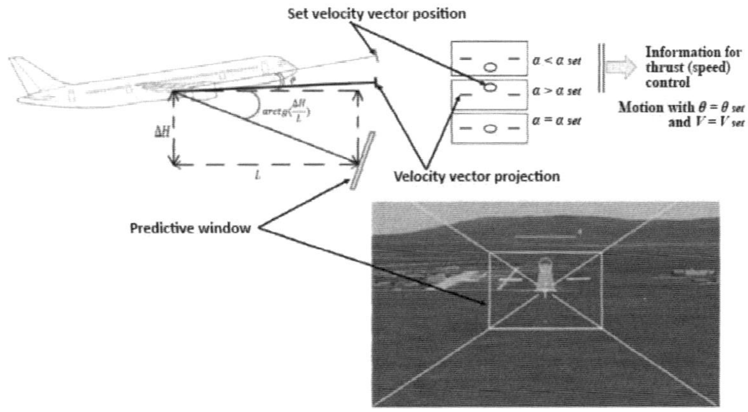

Figure 1. Predictive display.

In addition to the corridor, the display also shows a mark determining the set position of the velocity vector (see Figure 1), corresponding to the set landing speed, having the form of a fixed circle, which the pilot aims to align with the center of the predictive window using the control stick.

An important task during glidepath motion is holding the set airspeed. When performing a landing, the airspeed is close to the minimum, and the flight path angle control may become inverted under the influence of disturbances. Pulling the stick back will lead to a downward motion of the aircraft, because an increase in the angle of attack and induced drag divergence lead to a decrease in speed [4].

This peculiarity forces the pilot to simultaneously operate two controllers: the control wheel or control stick — for changing the pitch angle — and the thrust lever — for holding the airspeed and keeping a constant angle of attack.

At that, thrust control requires displaying on the screen the information on the difference between the trim and current angles of attack, proportional to the distance between the set velocity vector position and the current velocity vector position.

IOP Conf. Series: Materials Science and Engineering 1027 (2021) 012004 doi:10.1088/1757-899X/1027/1/012004

Thus, when deflecting the control stick, the pilot aims to align the set velocity vector position with the center of the predictive window, and, by operating the thrust lever, he aligns the velocity vector with the set velocity vector position mark.

This way of presenting flight information is applicable in the absence of continuous vertical wind gusts. In case of such gusts, if the velocity vector indicator goes below or above the set velocity vector value, and the airspeed matches the set speed, the pilot must align the velocity vector with the center of the predictive window and hold the set airspeed by referencing the speed indicator.

4. Conducting experimental research for evaluating the effectiveness of the information display system

The experimental research was carried out using the MAI flight simulator, equipped with a stereoscopic semi-panoramic visualization system, with the predictive display image being projected onto the external environment (Figure 2).

Figure 2. MAI flight simulator.

In the research, the following aspects were considered:

- the influence of the weight of a previously passing aircraft on the effects of a wake vortex encounter;
- the influence of the distance between a vortex and the runway;
- minimization of the effects of a wake vortex encounter when informed about it by the ground services.

During the investigation into the effects of "vortex generator" aircraft, two craft were examined: the A380 weighing 258 tons and the Legacy 450 weighing 14 tons. It is shown that increasing the weight of a "vortex generator" aircraft by 18 times leads to an average increase in the spread of glidepath deviations by 1.5 times. However, extensive use of the developed display by the pilot allows to perform piloting with standard deviations of the coordinates H and Z relative to the glidepath not exceeding 2 meters (Figure 3).

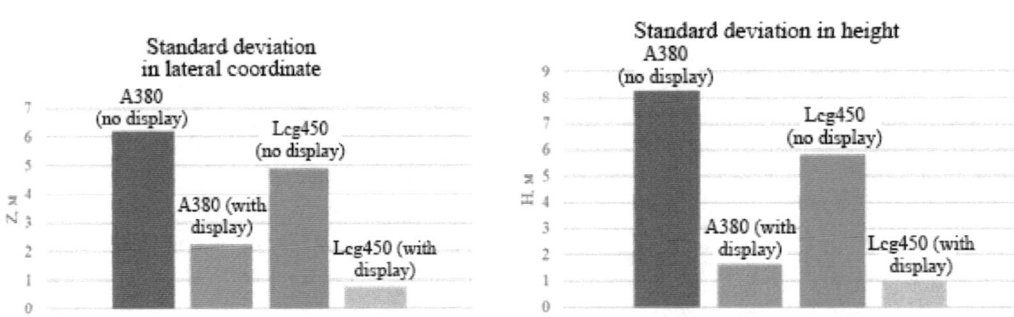

Figure 3. Glidepath tracking accuracy.

IOP Conf. Series: Materials Science and Engineering 1027 (2021) 012004 doi:10.1088/1757-899X/1027/1/012004

When researching the influence of the distance between a vortex and the runway, in addition to glidepath deviations, the probability of a pilot making the decision to land was estimated. For a pilot authorized to perform Category IIIA operations, the decision height for landing is 30 m. It is considered that a pilot can perform the landing if the deviation from the glidepath does not exceed 3.7 meters in height, 22 meters in the lateral coordinate, and 5 knots in speed.

The study considered two vortex positions: at a distance of 3 and 1.3 km from the runway threshold. The results of the research show that the probability of a pilot making the decision to land, calculated based on the above conditions with the vortex located at a distance of 3 km, is 55%, whereas in case of the vortex located at a distance of 1.3 km, it is merely 8%. Using the proposed indication allows to increase the probability of the pilot making the decision to land up to 95 and 46% with the vortex located at a distance of 3 and 1.3 km respectively. At that, the deviation from the glidepath will not exceed 3 meters in height and 5 meters in crossrange (Figure 4).

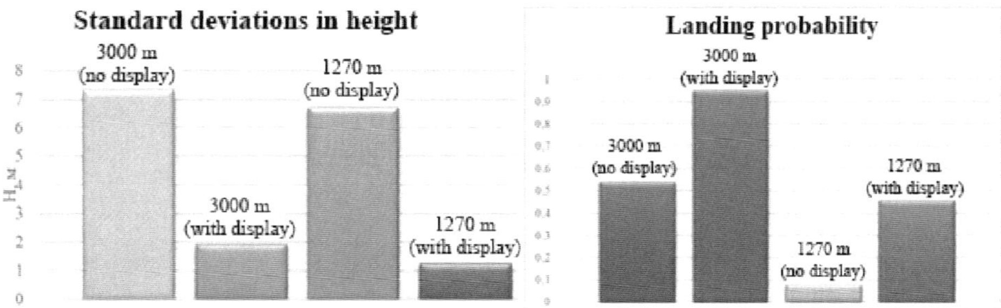

Figure 4. Glidepath height tracking accuracy and landing probability.

In addition, studies were carried out demonstrating the effectiveness of pilot warning by the controller about the presence of a wake vortex on the glidepath. If the pilot has been warned about a wake vortex, he increases the glidepath angle from 2.66 to 3 degrees and increases the approach speed by 5 knots. The studies were conducted with and without the predictive display. During research, the distance of the vortex relative to the runway was also changed.

It is shown that with the wake vortex located at a distance of 1.3 km from the runway threshold, complying with the ground services recommendations leads to an increase in the probability of the pilot making the decision to land by 4 times up to 32%. However, with the vortex located at a distance of 3 km, the probability of landing decreases from 55 to 45%, which is attributable to the complexity of tracking a steeper glidepath and high approach speed.

Using the predictive indication in conjunction with the warning allows to increase the probability of landing in the presence of a vortex close to the runway from 46 to 62 %. If the vortex is located at a distance of 3 km, the warning insignificantly reduces the probability of landing from 95 down to 90% due to high approach speed.

Regarding glidepath tracking accuracy, using the predictive display allows to ensure a glidepath deviation of no more than 4 meters in height and no more than 2 meters in crossrange (Figure 5).

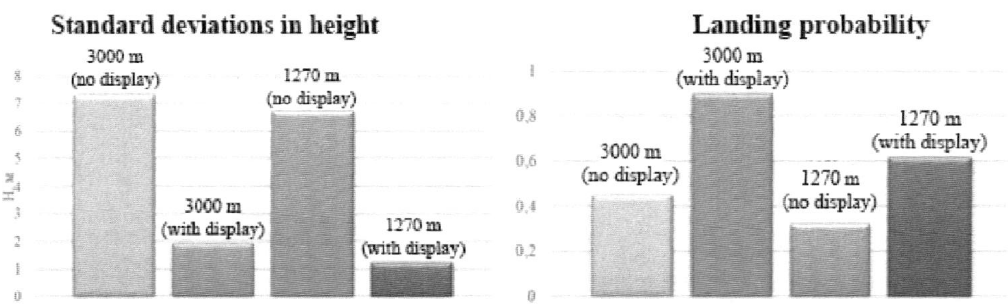

Figure 5. Glidepath height tracking accuracy and landing probability.

5. Conclusion

For the purpose of minimizing the negative effects of wake vortex encounters, a predictive information display system is proposed, whose parameters were selected by means of mathematical modeling of the pilot-aircraft system.

The experimental research evaluating the effectiveness of using the developed predictive information display system in a wake vortex encounter was carried out using a flight simulator. It is shown that using the indication allows to keep the spreads relative to the glidepath within 1.5–2 meters for varying intensity and vortex position relative to the runway and allows to increase the probability of the pilot performing the landing by 1.5–2.5 times on average.

Acknowledgements

The paper is prepared in the framework of the Program for the development of a world-class research center “Supersonic” in 2020–2025 founded by Russian Ministry of Science and Higher Education (Agreement 16 Nov 2020 № 075-15-2020-924).

References

[1] *Safety Report Edition* 2018, IATA 55
[2] Voevodin A V 2014 Applications of the Wake Vortex Mathematical Model *Trudy TsAGI* 2730.
[3] Efremov A V and Tjaglik M S 2011 *The Development of Perspective Displays for Highly Precise Tracking Tasks* Advances in Aerospace Guidance, Navigation and Control ed Holzapfel F and Theil S (Springer)
[4] Efremov A V, Zakharchenko V F, Ovcharenko V N, Sukhanov V L, Shelyukhin V F and Ustinov A S 2011 *Flight Dynamics* ed Bushgens G S (Moscow: Mashinostroyenie) 776

IOP Conf. Series: Materials Science and Engineering 1027 (2021) 012005 doi:10.1088/1757-899X/1027/1/012005

Numerical simulation of splashing landing gear on an amphibious aircraft air cushion

A V Fevralskikh

Moscow Aviation Institute (National Research University), 4 Volokolamskoe shosse, Moscow 125993, Russia

a.fevralskih@gmail.com

Abstract. The paper presents the basics of the methodology for modeling the spatter formation of landing gear on an amphibious aircraft hovercraft based on the numerical solution of the Navier-Stokes equations for a two-phase incompressible flow with a free interface (VoF approach) and the disperse medium motion equations (DPM approach). The results of numerical simulation of the water jet collapse in the transverse air flow in comparison with the experimental results are given, as well as recommendations and results of application of numerical simulation in forecasting the air cushion splash formation. The results of the simulation are in good correspondence with the results of the experiments published earlier, which demonstrates the prospects of applying the technique in the development and design of new landing gear layout solutions on an amphibious air cushion aircraft.

1. Introduction

The advantages of hovercrafts are related to the properties of amphibious aircraft, which allow for takeoff and landing on water or snow surface, unequipped sandy shore or shallow water, as well as to approach the place of passengers' disembarkation, using the bearing properties of the air cushion. The capabilities of amphibious vehicles allow to consider them as an actual component of an off-road module to ensure the interconnection of the Russian Federation territory when forming complex scientific and technical projects [1]. At the same time, the operation of amphibious aircraft involves the risk of icing on takeoff, just as it is typical for hovercrafts, because the flow of air from under the air cushion fence provokes an intense spray. Icing of an amphibious aircraft can lead to changes in wing bearing properties, increased takeoff weight, damage to propellers and a number of other negative effects.

At the same time, the prediction of splashability and icing of an amphibious aircraft on takeoff is a nontrivial task, to solve which, in general, it is necessary to take into account the nature of the airflow and droplets from the air cushion, and the external aerodynamic flow of the fuselage, and the interaction of droplets with the wing surface and the fuselage, and the interaction of splashes with the external aerodynamic flow. In order to study the aerohydrodynamics of designed amphibians, large scale models are traditionally tested in hydro canals [2–4]. Despite a number of limitations associated with the effect of scaling, hydrochannel testing can determine with sufficient reliability for design purposes characteristics such as the relative resistance of an air cushion and the running differential at different Froude numbers. At the same time, the numerical simulation of aerohydrodynamics (CFD — computational fluid dynamics), based on the Navier-Stokes equations solution by reference volume

IOP Conf. Series: Materials Science and Engineering 1027 (2021) 012005 doi:10.1088/1757-899X/1027/1/012005

method, with the use of Euler's Volume of Fluid (VoF) approach, is increasingly being used for such tasks [5, 6]. CFD modeling tools have been implemented in CAD software products, which determines their availability and ease of use, and the validation results confirm the effectiveness of CFD for amphibian design tasks.

At the same time, the application of the control volume method together with VoF approach has such limitations as iterative process instability, high requirements to the quality and quantity of the computational grid elements (with the corresponding computational resource costs). To predict splash dynamics using the VoF approach, generally speaking, it is desirable to know in advance the approximate distribution of droplet sizes, on the basis of which you can set the initial distribution of control volume sizes and set up the grid model. These requirements encourage the development of new methods of numerical simulation of aerohydrodynamics and spatter formation, which would combine such qualities as reliability and robustness, the possibility of application in automated design. For development of such methods it is supposed to consider Lagrangian and pseudo-Lagrangian approaches based on representation of droplets as discrete particles.

In the present work the possibilities of the computational method of aero-hydrodynamics simulation of VoF-to-DPM for the development of the method of numerical simulation of the air-cushion chassis splash formation are investigated. The VoF-to-DPM method is based on the Euler fluid representation with the possibility of converting its individual fragments into the Lagrangian Discrete Phase Method, i.e. it is a hybrid approach. The paper presents the basics of the method of modeling the spray formation of an air cushion using the DPM method, presents the results of numerical simulation of the water jet collapse in the transverse air flow in comparison with the results of experiments, as well as recommendations and results of the method in modeling the spray formation of an air cushion.

2. Fundamentals of DPM method

DPM (Discrete Phase Method) refers to the Euler-Lagrangian approaches, where the dynamics of a solid medium are modeled using the reference volume method using Navier-Stokes equations and the motion of discrete particles is described according to the law of pulse change (1).

$$m_p \frac{d\vec{u}_p}{dt} = m_p \frac{\vec{u} - \vec{u}_p}{\tau_r} + m_p \frac{\vec{g}(\rho_p - \rho)}{\rho_p} + \vec{F}. \tag{1}$$

In the equation (1): m_p — particle mass, $\vec{u}_p$ — particle velocity, $\vec{u}$ — rate of flow, $\vec{g}$ — free-fall acceleration, ρ и ρ_p — respectively the density of the solid medium and the density of the particle, $\vec{F}$ — additional forces influencing the motion of a particle, τ_r — time of relaxation (2).

$$\tau_r = \frac{\rho_p d_p^2}{18\mu} \frac{24}{C_d Re}. \tag{2}$$

Relaxation time τ_r is determined by the molecular viscosity of the solid medium μ, particle size d_p, particle resistance coefficient C_d and Reynolds' number Re by the velocity of the particle relatively solid medium.

Approach described by the equations (1) and (2) in general, it is quite universal, and can be complemented by relations describing the influence of particles on the heat transfer, turbulence and other characteristics of the solid medium — on the one hand, and the relations that determine the forces acting on the particle side of the solid medium — on the other hand. Separately, equations can be introduced to describe the dynamics of particles in terms of degrees of freedom of rotation. With this approach, it is assumed that spatter (droplets) have a spherical shape, the drop size does not exceed 10% of the characteristic size of the reference volume, and the processes of flow on the surface of the drop do not have a noticeable effect on the motion characteristics. In this paper, to determine the resistance coefficient C_d approach was taken *Dynamic Drag Model* (3), implemented in Ansys Fluent software, which allows to take into account the change in particle shape during drop destruction [7].

$$C_{d,sphere} = \begin{cases} 0.424 & Re > 1000, \\ \dfrac{24}{Re}\left(1 + \dfrac{1}{6} Re^{2/3}\right) & Re \le 1000, \end{cases}$$

$$C_d = C_{d,sphere}(1 + 2.632y),$$

$$\frac{d^2 y}{dt^2} = \frac{C_F}{C_b}\frac{\rho_g}{\rho_l}\frac{u^2}{r^2} - \frac{C_k \sigma}{\rho_l r^3} y - \frac{C_d \mu_l}{\rho_l r^2}\frac{dy}{dt}. \qquad (3)$$

Model represented by a system of equations (3), is based on Taylor's analogy [8]. In the system of equations (3) y — drop deformation value (deviations from the spherical shape), r — nondeformant drop radius. Model factors $C_b = 1/2$, $C_F = 1/3$, $C_d = 5$, $C_k = 8$ are determined by the results of experiments.

Madabhushi model with Ansys Fluent settings offered by default is used for modeling decomposition of liquid fragments (Figure 1).

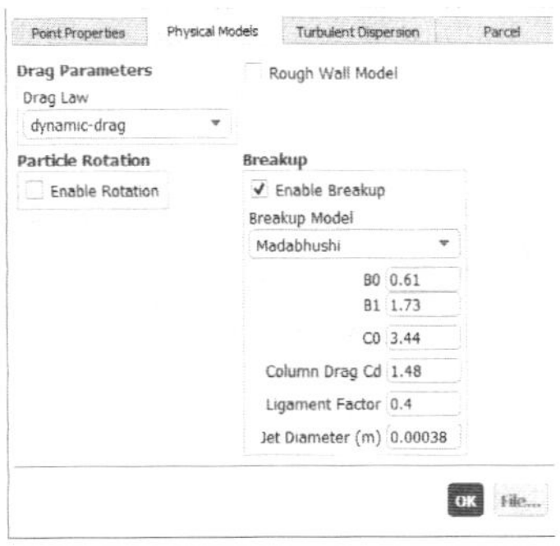

Figure 1. Settings of the decomposition model of liquid fragments in Ansys Fluent.

3. Simulation of a water jet collapse in a transverse air flow

The water jet spray in the transverse air flow is simulated (Figure 5). The characteristics of the modeled flow are given in Table 1.

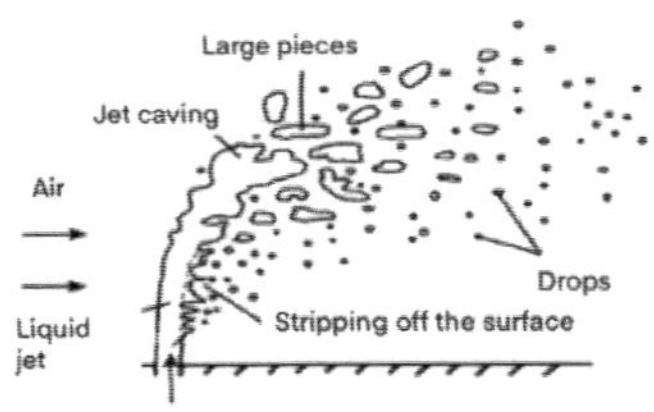

Figure 2. Modeling the collapse of the water jet in the transverse air flow.

Table 1. Characteristics of the modeled flow.

Water Density ρ_l, kg/m³	Air density ρ_a, kg/m³	Weber Number We	Dynamic pressure ratio $q = \dfrac{\rho_l V_l^2}{\rho_a V_a^2}$	Jet diameter in injection area, mm	Surface tension factor σ, N/m
998.2	1.225	248	7.1	0.381	0.0719

IOP Conf. Series: Materials Science and Engineering 1027 (2021) 012005 doi:10.1088/1757-899X/1027/1/012005

The joint turbulent flow of two incompressible liquids — air and water — with free interface of variable topology is investigated. It is accepted that turbulization of the aerodynamic flow caused by the appearance, destruction and movement of droplets under these flow conditions does not have a noticeable effect on the velocity distribution. Numerical modeling of control volumes using DPM approach was performed in Ansys Fluent software. A polyhedron mesh model with the number of control volumes of 30108 units was used, a turbulence model SST k-ω and a turbulence damping option were connected to close the system of Navier-Stokes averaged Reynolds equations [6]. Figure 3 shows the input (for air) and the lower boundary of the simulation area with a splash cloud obtained from the simulation results.

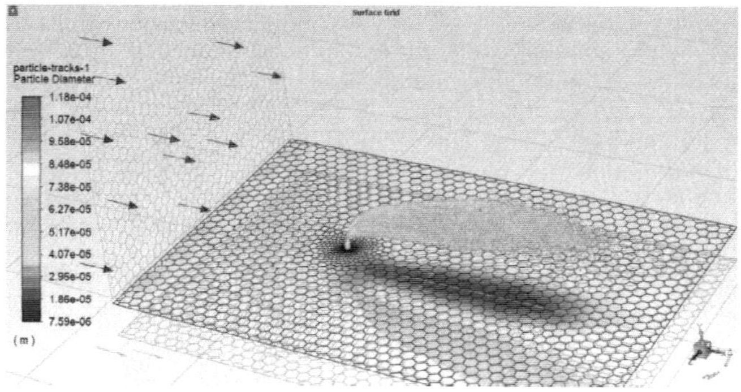

Figure 3. Displaying the boundaries and results of spray modeling in Ansys Fluent.

At the input boundary of the modeling area, the boundary condition of aerodynamic flow velocity was set, at the output boundary — pressure condition, at the upper, lower and side walls — impermeability condition with slippage. Also, at the lower boundary is given a source of injection of spray — water as a discrete phase, according to the conditions given in Table 1.

The computation time was about 18 hours on a computer with 4 physical cores of 3.4 GHz Intel Core i7-6700 processor, 64 GB of RAM and an NVIDIA Quadro K620 graphics card.

The results of the simulation of the particle trajectory were compared with the correlation of the results of experimental studies [9–14].

$$\frac{y}{D} = 1.55q^{0.53} \ln\left(1 + 1.66\frac{z}{D}\right). \tag{4}$$

In relation (4) y — vertical co-ordinate of the particle, z — co-ordinate of the particle in the direction of aerodynamic flow, calculated from the point of injection (injection) of water jet, D — jet diameter at the injection point. The mean square deviation of the obtained correlation is 1.83 units.

The results of DPM numerical modeling performed in Ansys Fluent compared to the results of the experiment in the form of discrete phase coordinate distribution (spatter/droplets) are shown in the figure below 4.

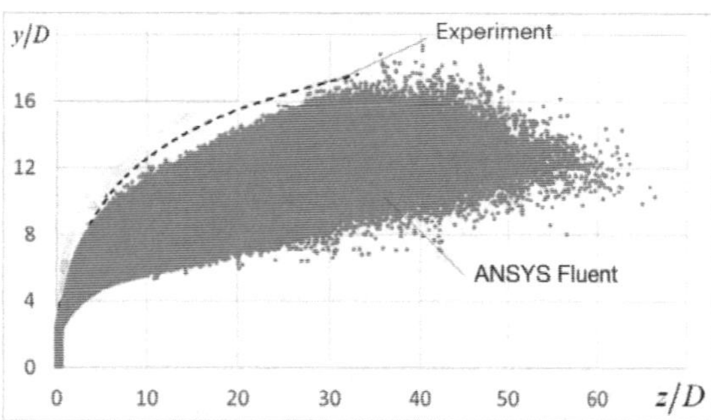

Figure 4. Comparison of the results of DPM water jet decay simulation with the results of experiments.

The results of the numerical simulation are in good agreement with the results of the correlation of experimental data in the area of jet injection and for most DPM particles in the area of jet disintegration. However, the default settings of the Madabhushi model offered in Ansys Fluent seem to be subject to additional calibration to achieve a better match with the experimental results.

4. Numerical simulation of hovercraft spraying

Numerical simulation of the spray formation of the amphibious aircraft air cushion section, shown in Figure 5, was performed. The graphical image of the calculated area is presented in Figure 6.

Figure 5. Hovercraft with chassis on a hover cushion on takeoff mode from water.

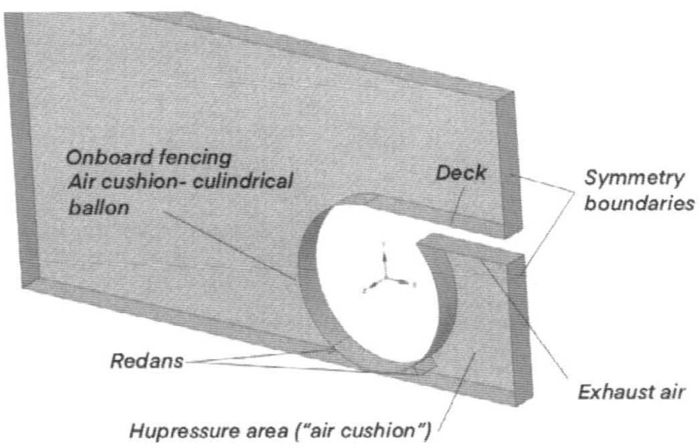

Figure 6. Scheme of the calculation area of air cushion section modeling.

The results of modeling are given in Figure 7 in comparison with the results of experiments and calculations according to the empirical model [15] at different values of air flow rate and initial draft of the vessel. Quantitative analysis of modeling results is presented as a dependency of dimensionless volumetric spray rate $q_m^* = q_m / Q_a$ (q_m — volumetric spray rate, m^3/s; Q_a — volumetric airflow rate, m^3/s) against the criterion U (5).

$$U = ln\left(10^5 \frac{Q_a^* r_m}{H_n}\right),$$
$$Q_a^* = \frac{Q_a}{H_n \sqrt{2P/\rho}}. \tag{5}$$

In relation (5) H_n — the depth of the air cushion trough, m; P — average air cushion pressure, Pa; ρ — air density, kg/m^3; r_m — average median drop radius, determined by the results of numerical modeling.

The computation time for each calculation case using the same computer as in the jet disintegration simulation was about 36 hours.

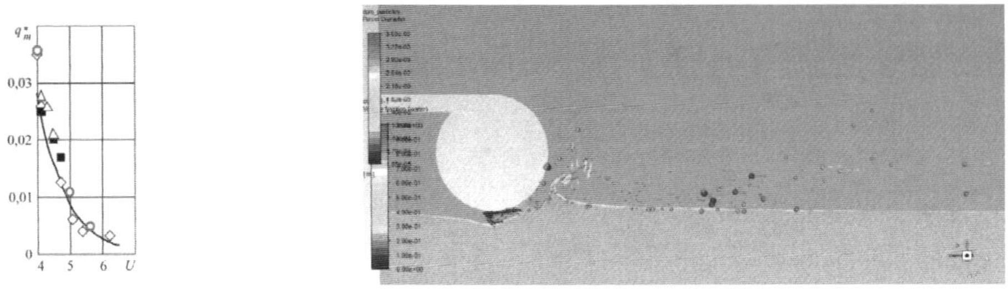

(a) (b)

Figure 7. Results of numerical simulation of the air cushion spray using the VoF-to-DPM method: (a) — comparison of the values of dimensionless spray flow at different values of the criterion obtained during CFD modeling (round markers) with the data of experiments and empirical model; (b) — visualization of free surface and spray near the air cushion body.

The data in Figure 7 show that the application of VoF-to-DPM method in CFD modeling allows to reliably determine the amount of water carried as a spray from the air cushion area. In this case, there is no modeling is carried out in a full-size setting and, thus, it is possible to avoid the theoretical assumptions accepted when setting the model experiments in experimental pools. In addition, when performing this kind of calculations, it is not necessary to specify the depth of the trough, the pressure value and accept M.S. Volynsky's condition as the initial data, as proposed in the method of [15]. In addition, if sufficient computational resources are available, the application of this technique can be extended to the case of a complete aircraft layout, thus analyzing the interaction of spray with propellers, sprayability of the fuselage and other effects. However, despite the high level of automation of the described method of modeling spatter formation, its requirements to computational resources remain quite high, and in the future, it seems expedient to consider the use of meshless methods of modeling disperse hydrodynamics.

5. Conclusion

The basics of the method of numerical simulation of splash formation of landing gear on the air cushion of amphibious aircraft are described. The technique is based on the numerical solution of the equations of viscous turbulent flow of incompressible fluid by the method of control volumes together with the equations of motion of disperse medium. The VoF-to-DPM method used in this case, implemented in Ansys Fluent software, allows to take into account in one statement of the computational experiment the deformation of the free water surface and the dynamics of splashes represented by discrete particles. The results of the method verification in the course of numerical simulation of water jet disintegration in the air current demonstrate good agreement with the experimental results. The reliability of the results of the air cushion section modeling is confirmed by comparison with the results of the empirical model calculations. Among advantages of the stated technique it is necessary to carry high automatization, possibility of application together with software complexes of the automated designing, and also possibility to leave a number of theoretical assumptions and to apply technique in the decision of problems of interaction of full configuration of an aircraft with splashes.

References

[1] Pogosyan M A, Strelets D Y and Vladimirova V G 2019 Territorial Connectivity of the Russian Federation: From the Statement of Complex Problems to Drawing up Integrated Scientific and Technical Projects *Herald of the Russian Academy of Sciences* **89** (2) 179–84

[2] Fevralskikh A V 2017 *Development of methods for designing the aero-hydrodynamic arrangement of an amphibious aircraft on an air cushion with aerodynamic unloading on the basis of numerical modeling* Cand. (Nizhny Novgorod: R.E. Alekseev NGTU) 175 p

[3] Diachenko N V 2013 *Spattering of amphibious air-cushion vessels* Dissertation of Doctor of Technical Sciences (St. Petersburg State Maritime Technical University) 301 p
[4] Demeshko G V 1992 *Design of ships. Amphibious FPVs Textbook* Vol 1 (St. Petersburg: Shipbuilding) 269 p
[5] Kal'yasov P S, Fevralskikh A V and Shabarov V V 2014 Mathematical modeling of the amphibian vessel aero-hydrodynamics on an air cushion *J. Problems of Strength and Plasticity* **76** (3) 263–8
[6] Gramuzov E M and Fevralskikh A V 2018 Application of the turbulence damping method for calculation of the water surface shape in the task of determining the resistance of an air cushion vessel *Sci. J. Marine intellectual technologies* **4** (42) 242–9
[7] Liu A B, Mather D and Reitz R D 1993 *Modeling the Effects of Drop Drag and Breakup on Fuel Sprays* SAE Technical Paper 930072
[8] Taylor G I 1963 The Shape and Acceleration of a Drop in a High-Speed Air Stream Technical Report *Scientific Papers* ed G K Batchelor
[9] Tambe S, Jeng S-M, Mongia H and Hsiao G 2005 Liquid Jets in Subsonic Crossflow *43rd AIAA Aerospace Sciences Meeting and Exhibit*. doi:10.2514/6.2005-731
[10] Becker J and Hassa C 2002 Breakup and Atomization of a Kerosene Jet in Crossflow at Elevated Pressure *J. Atomization and Sprays* **11** 49–67
[11] Inamura T, Nagai N, Hirai T and Asano H 1991 Disintegration Phenomena of Metalized Slurry Fuel Jets in High Speed Air Stream *Proc. of the 5th Int. Conf. on Liquid Atomization and Spray Systems* Gaithersberg, MD USA p 839
[12] Chen T H, Smith C R and Schommer D G 1993 *Multi-Zone Behavior of Transverse Liquid Jet in High-Speed Flow* AIAA Paper 93–0453
[13] Wu P-K, Kirkendall K A, Fuller R P and Nejad A S 1997 Breakup Processes of Liquid Jets in Subsonic Crossflows *J. of Propulsion and Power* **13** (1) January-February 64–72
[14] Yates C L 1971 *Liquid Injection into Supersonic Airstreams* AIAA Paper 71–724
[15] Anosov V N and Diachenko N V 2009 Calculation of the amount of water carried into the atmosphere by an air jet from an air cushion *Proc. of the Central Res. Inst. of Acad.* A N Krylov **49** (333) p 123

Technology for aerodynamic wake characteristics measurement behind an aircraft on a rail-guided track

I G Golovnev[1], V V Vyshinsky[2], O V Sokolov[1], K V Lapshin[1] and P A Mukhachev[3]

[1] State Research Institute of Aviation Systems, 7 Viktorenko, Moscow 125319, Russia
[2] Zhukovsky Central Aerohydrodynamic Institute (TsAGI), 1 Zhukovsky street, Zhukovsky 140180, Moscow Region, Russia
[3] Skolkovo Institute of Science and Technology (SkolTech), Moscow 121205, Russia

golovnev@gosniias.ru

Abstract. A framework for vehicle wake vortexes experimental investigation on a rail-guided track in sub- and supersonic flows is presented. Basic requirements for measurement system and results treatment are determined based on numerical modeling and wind-tunnel tests.

1. Introduction

Coherent vortex structures in the atmosphere are represented with strong vortexes that are stable enough for their lifetime to be larger than elementary volume rotation period around instant axis. At the same time, their nature is stochastic and unpredictable. Classical examples are vortices behind aircraft (Figure 1a), vortex wakes behind air-carriers (Figure 1b), wind vortex structures in a mountain landscape (Figure 1c).

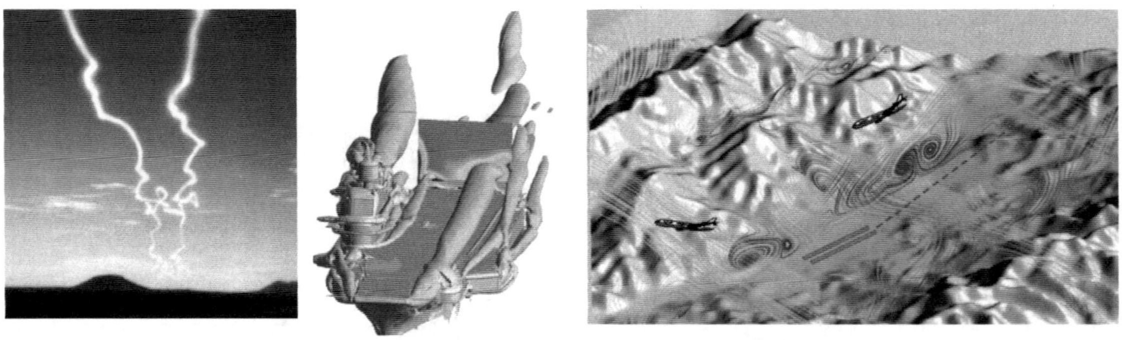

(a) (b) (c)

Figure 1. Coherent vortex structures examples: (a) aircraft vortex wake, (b) aircraft carrier wake, (c) wind vortex structures under mountain landscape.

Entry into the coherent vortex structures can lead to unexpected abrupt bank angle changes, which is especially dangerous at low altitudes, autopilot deactivation, engine cut-out, crew errors, hard landing out the runway (RW) and others. Passing across the vortex is dangerous in terms of dynamical loadings and structural damage. Vortex wake strength determines the delay between aircraft landing on the same RW and take-off and eventually the overall throughput of an airport. ICAO recommends to set safe distance between two consequent aircraft depending on their classes. However, sometimes accidents occur even if all the recommendations and air traffic controller (ATC) commands are carried out [1].

There are many factors that influence on the evolution and decay of wake vortexes, such as wind in terms of vortex transfer (in particular, wind shear that carries vorticity), stratification (vortex cluster descending against its buoyancy increases vortex dissipation rate), atmospheric turbulence. Due to mutual induction, the vortexes go down, earth influence causes them to spread out and raise themselves. Side wind carries them, keeping the upwind one stay above RW for some time, which is the worst case for flight safety. The influence of random wind gusts leads to long-wave instability onset. Direct numerical simulation in the framework of the boundary-value problem for filtered Navier-Stokes equations (Large Eddy Simulation) with subgreed turbulence model (LES) reproduces small-scale as well as long wave instability development [2].

Current simplified models doesn't account for these factors or do that inaccurately. Due to that fact the efforts to make vortex safety system based on vortex wake evolution prediction are not likely to be successful. It is required to see or sense wake vortexes in time. The former is possible with lidars, IR-cameras or meteoradars. The latter is possible with onboard system detecting the trajectory deviation from the trajectory of controlled flight in the calm atmosphere. In both cases the problem of image recognition arises, for which machine learning techniques can be employed.

In order to provide data about atmospheric conditions in the area of the runway, missing data have to be generated based on limited amount of information from sensors. For this task, generative artificial neural networks (ANN) can be employed. There is a good progress in system identification and modeling of unknown nonlinear processes using reccurent ANN. As soon as the flow pattern (atmospheric conditions) in the area of interest is known, the decision has to be made on whether to allow takeoff or landing. It is generally quite hard for human operator to make such a decision based on raw flow pattern, thus, decision support is needed. The most important function of that system is to recognize vortexes, determine their size and strength, and to provide relevant information to ATC. For the task of pattern recognition, convolutional neural networks have proved their efficiency and could be successfully used for the task.

In this work, methodological questions of using a rail track for experimental measurement of vortex wake parameters at sub- and supersonic conditions for flight safety purposes are discussed. Experimental data received will be used for the verification and validation of the computer's codes used for the aircraft vortex wakes simulation.

2. Aircraft vortex wake

Wake formation was demonstrated for A380 aircraft during final approach (flight on glidepath, mass 380 t, velocity 70 m/s, wing span 83 m). Parameters of the atmosphere: temperature 273K, pressure 70121.2 Pa [3]. Angle of attack 13.5 degrees. Core circulation 420.35 m^2/s, core radius 3.39 m.

On figure 2 wake parameters in control section 3 seconds after the aircraft crossed it are presented. Three distinct pairs of vortexes can be observed. Contour corresponds to the position of the wake generating aircraft during its traversal through the cross-section.

The modeling is conducted according to [3–7]. The wake is represented by the structure of several pairs of vortexes with opposite directions of rotation. One pair is generated by wingtips, two pairs by the flaps, and one pair is due to the fuselage. External flap vortexes has the same direction as those by the wingtips and could be stronger.

Unexpected wake vortex entering won't allow the pilot to have a soft landing on RW. Timely notification can alleviate the consequences (figure 3).

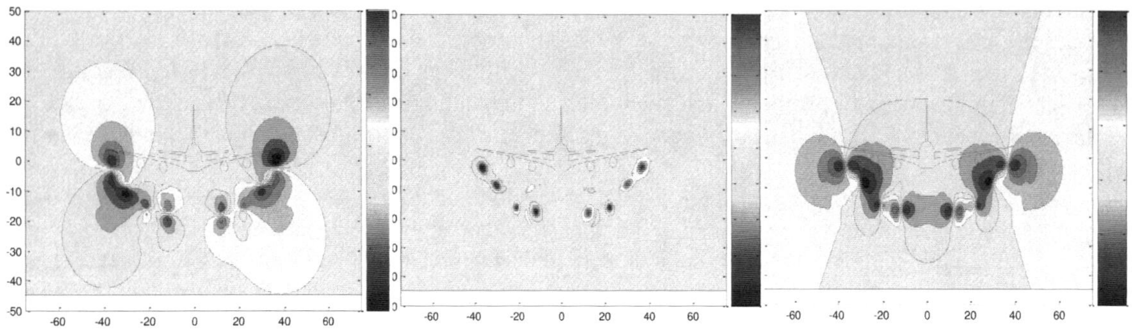

Figure 2. Fields of horizontal and vertical velocity components, longitudinal vorticity component and aircraft contour 3 seconds (210 meters) behind the aircraft.

In [4–7], the possibility of constructing an early warning system for a pilot when an aircraft - subscriber enters the vortex wake from an aircraft — generator based on the use of an on-board air signal system and cross flow air sensors was considered.

Numerically verified algorithm was validated in T-103 TsAGI's wind-tunnel tests. Near wakes behind two aircraft models were simulated, in which the distinct coherent structures are not present yet.

Algorithm generated the alarm signal if the all three pair of sensors reach agreement in center of vorticity prediction in the framework criteria used.

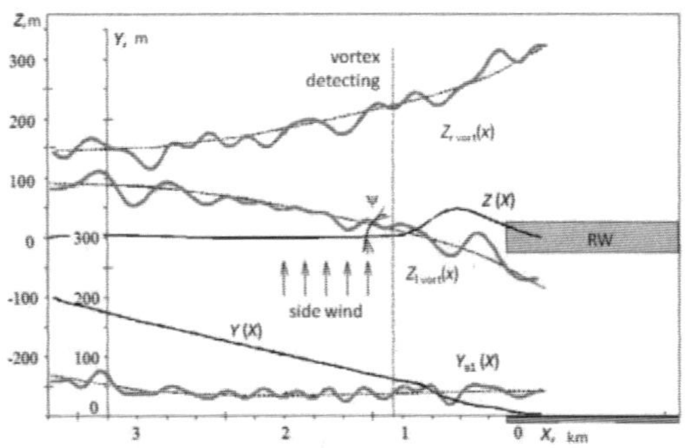

Figure 3. Pilot-in-the-loop simulation the aircraft entering the left vortex behind the preceding aircraft during flight on glidepath.

The results wake vortex scanning behind Il-106 aircraft (figure 4) in TsAGI's T-103 wind tunnel with elliptic test section axes 4 × 2.33 m, at flow speed V = 50 m/s are represented in figure 5.

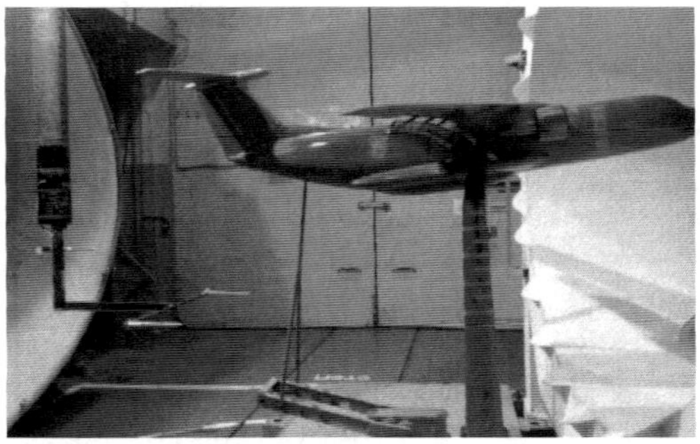

Figure 4. Il-106 model and three 6-point cross flow sensors, which act as probe aircraft.

Measured vertical and horizontal velocity components and static pressure fields in the control section are shown on figure 5. Pressure difference in control section that can be used to detect vortex structure core is 20 Pa. Working area boundaries as well as contour of an aircraft generating the wake are also shown. Origin of coordinates is situated in the center of the model (in the intersection of

IOP Conf. Series: Materials Science and Engineering 1027 (2021) 012006 doi:10.1088/1757-899X/1027/1/012006

fuselage construction horizontal line and vertical plane crossing the front mounts of the model support system).

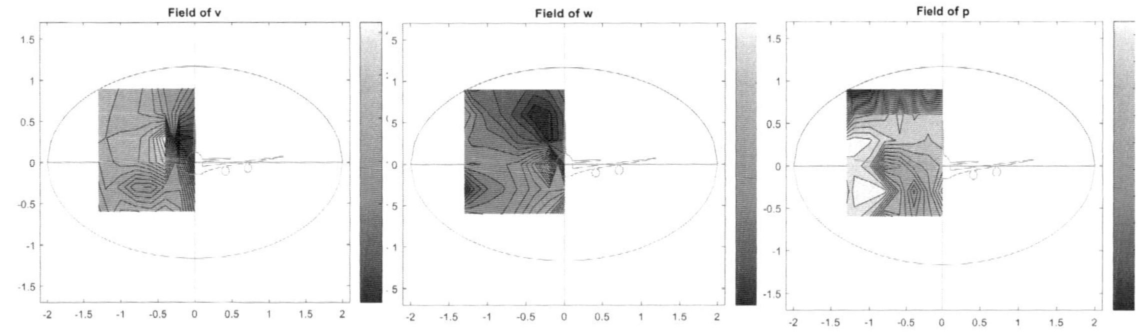

Figure 5. Vertical and horizontal velocity distributions and pressure distribution in control cross-section.

Proposed and verified during numerical modeling and wind tunnel tests algorithm can be proposed for experimental verification in direct motion.

3. Rocket-powered rail track as a mean for wake vortex experimental modeling in straight motion.

Validation of onboard real-time expert system for wake vortex detection has to be conducted in straight flow. To do that high-speed track can be employed, including rocket-powered track. Principal trolley sketch is shown on figure 6. The experiment has to be conducted in 2 stages.

In the first stage wing model (8) is set on the car (4) via support system (5). The car is mounted to the rails (10) stopped on them by rail shoes (12). Behind the wing model the comb with pressure and velocity sensors (3) is installed for continuous flow measurements in the near wake. In front of car (4) pressure sensors are installed according to figure 6. Behind the car with the wing model, rocket-powered car (1) with solid fuel rocket engine rigidly connected with it is installed. On rocket-powered car the comb with pressure and velocity sensors (3) is set. Axes of two wake combs mounted on car with model and on car with engine should be aligned.

In the second stage, the rocket engine is turned on and both cars drive along the track. At a certain point A (see figure 7), rocket-powered car is uncoupled and in the second stage car with the model moves freely before full stop. Vortex wake parameters measurement behind the wing model is continuous throughout the experiment.

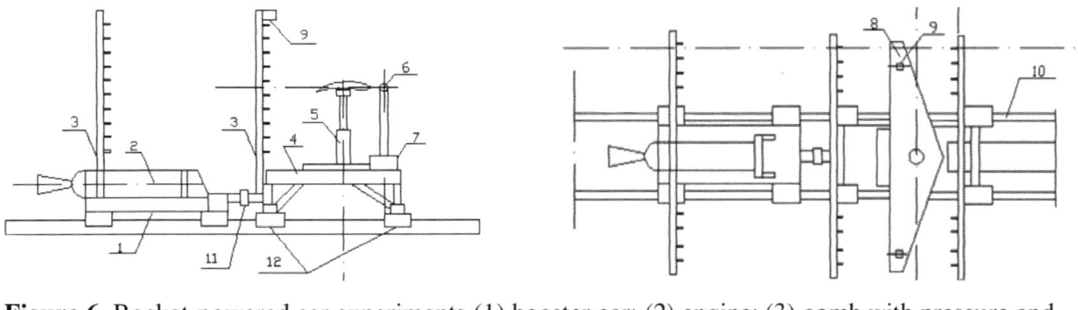

Figure 6. Rocket-powered car experiments (1) booster car; (2) engine; (3) comb with pressure and velocity sensors; (4) car with a model; (5) support system; (6) visualization system; (7) smoke generator; (8) wing model; (9) camera and/or IR-camera; (10) rail track; (11) car connection; (12) rail shoes.

Civil aircraft flight velocity ranging from 70 to 300 m/s can be realized on 2500 m length track: the acceleration to 300 m/s will take about 750 m, deceleration from 300 m/s to 70 m/s on 1000 m track will last for–3-10 seconds, and deceleration from 70 to 0 m/s will take 750 m length. Measured in forward flow vortex characteristics can be employed to validate numerical methods.

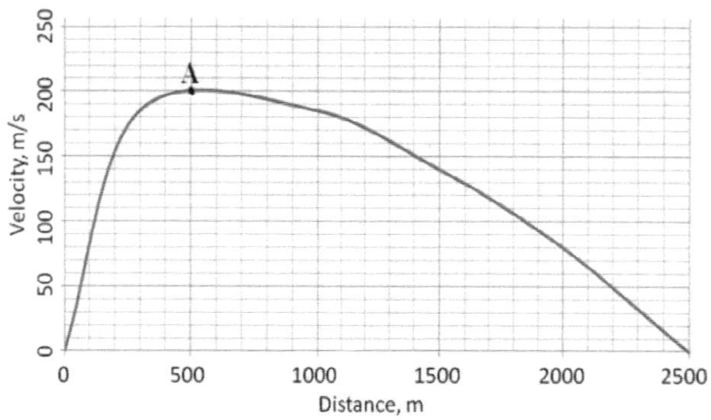

Figure 7. Graph of the model movement along the rocket track.

4. Conclusion

Technology foundations are proposed for modeling evolution and decay of vortex wake behind an aircraft in forward motion on rocket-powered trolley in short- and long-range vortex wake at sub-and super-sonic speeds. The results of the experiment can be employed for validation of numerical schemes, engineering practices, as well as for onboard vortex detecting and avoidance systems.

One of the possibilities to develop the system is to apply it for flight safety issues, connected to in-flight refueling (automatic refueling system), air carrier landing, close formation flight, mountain terrain airport landing, flight and other operations in mountains landscape and near high rise buildings (in orographic turbulence conditions), clear-sky turbulence detection during cruise flight.

References

[1] Gerz T, Holzaephel F and Darracq D 2001 *Aircraft Wake Vortices. WakeNet Position Paper*
[2] Bosnyakov I S and Soudakov G G 2015 Verification of the engineering model of the destruction of the vortex wake behind an airplane using the method of large eddy simulation *Trudy, MIPT* **7** 2 83–98
[3] Vyshinsky V V, Zamiyatin A N and Soudakov G G 2006 Theoretical and experimental study of the evolution of the vortex wake behind an airplane flying in the atmospheric boundary layer *Technika vozdushnogo flota* **3–4** 25–38
[4] Vyshinsky V V, Sviridenko Yu N, Golovnev I G and Lapshin K V 2019 On-board vortex detecting and avoidance system *Trudy GosNIIAS. Ser. Voprosy avioniky* **1** (41) 35–49
[5] Golovnev I G, Vyshinsky V V, Kulchak F M, Lapshin K V, Zimin R Yu and Lubaeva D S 2019 The on-board operational-advisory expert system that provides warning to the crew of an aircraft about entering a vortex trace from another aircraft *Navigation, guidance and control of aircraft Abstracts of the Fourth All-Russian Scientific and Technical Conf.* p 320
[6] Golovnev I G, Lapshin K V and Falkov E Ya 2017 *A bench for determining the aerodynamic characteristics of a vortex of a wing model of an aircraft on a rocket track* Utility model Patent RUS 178659 17.07.2017
[7] Golovnev I G, Vyshinsky V V, Zhelannicov A I and Lapshin K V 2018 Principles of constructing an on-board early warning system for a pilot about entering a vortex wake from another aircraft *Nauchnyi vestnic MGTU GA* **21**(4) 84–95

IOP Conf. Series: Materials Science and Engineering 1027 (2021) 012007 doi:10.1088/1757-899X/1027/1/012007

About the equilibrium of airship in the parking lot

V T Grumondz, N I Morozov and Moung Htang Om

Moscow Aviation Institute (National Research University), 4 Volokolamskoe Shosse, Moscow 125993, Russia

v.grumondz@gmail.com

Abstract. The paper discusses the questions of constructing a mathematical model of the equilibrium of an airship in a parking lot with a two-cable parking scheme and also the problem of calculating forces acting on the elements of the cable system that holds the airship.

1. Introduction

Nowadays, many countries are working on the design of airships, analysis of their flight characteristics and research of their behavior in various parts of the trajectory [1–27]. One of the most important problems in this case is the problem of modeling the behavior of the airship in the parking lot, especially in the case of parking using a cable system, since the reliable equilibrium of the airship in the parking lot determines the safety of its operation as a whole. One may say, that the cable parking of the airship in the parking lot might be the most stressful and dangerous in all the airship flight operation.

In this regard, it is necessary to consider the construction of a mathematical model of the equilibrium of the airship in the parking lot first. This question is also important because in the airship equilibrium position it is possible to measure the aerodynamic forces and moments acting on it, just as it is done when conducting experiments in a wind tunnel. For this purpose, it is enough to provide each stretch with a dynamometer that can record the tension force of the cable, and then calculate the total force acting on the airship in one direction or another. In this case it is important that the experiment is carried out not on reduced models, as it is done in a wind tunnel, but on a real airship, and therefore it is not necessary to recalculate the aerodynamic characteristics from the model to nature. The results obtained in this paper are sufficient to solve this problem. In addition, they can be used to estimate the required strength of cables.

2. Basic assumption

Let's assume that the airship, while parked, is held in a stationary position by cables attached to some devices located on the ground (winches, etc.). The dynamic diagram of this position for a two-cable scheme is shown in Figure 1.

In Figure 1 A, B, a, b — the mounting point of the cables on ground and on airship body, respectively; l_A, l_B — cable lengths; $\bar{N}^A, \bar{N}^B$ — the reaction forces of the bonds; $\bar{A}, \bar{G}$ — aerostatic force and the weight force; O — center of volume; O_1 — center of mass; $O_0 x_g y_g$ — earth coordinate system fixed in space; Oxy — the coordinate system associated with the airship, the origin of which is located in the center of the volume. The axis Ox is directed along the longitudinal axis of symmetry of the hull towards its bow tip; $\overline{W}$ — wind speed vector at the earth's surface. We assume that $W = const$, $\overline{W} \| O_0 x_g$, and that all forces acting on the airship are located in a plane (a flat problem). The elasticity of the airship hull and its deformations will be ignored. Accounting for these deformations is an important task that requires separate consideration. In particular, the matrix of attached masses of a deformable airship will have a significantly more complex appearance than in the case of a rigid hull.

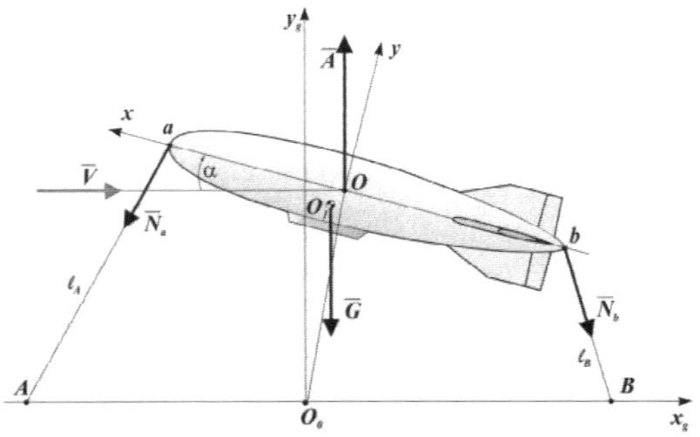

Figure 1. Dynamic diagram of airship arrangement in two-cable mounting scheme.

3. Nonlinear mathematical model of equilibrium for two-cable mounting scheme

In this case, the main problems are the equilibrium position of the airship, the cable mounting scheme, the estimation of forces arising in the mounting points, and the previously mentioned problem of experimental obtaining of the aerodynamic characteristics of the airship. The task of determining the mounting scheme is mainly understood as the choice of the position of points A, B and the length of the cables. Everywhere below, it is assumed that in this statement, it is possible to ignore the slackness of the cables and imagine them located along a straight line.

To determine the equilibrium conditions of a non-free rigid body, it will be considered that this body as free, replacing the bonds imposed on it with unknown reaction forces, the action of which on the body is equivalent to the action of bonds.

In the plane case under consideration, the forces and moments acting on the airship when the bonds are released are as follows: $\overline{N}^A, \overline{N}^B$— bond reaction forces; $\overline{G}$— weight applied at the center of mass, which generally does not coincide with the center of volume; $\overline{A}$— aerostatic force applied at the center of the volume; $\overline{R} = (R_x, R_y, R_z)$— the aerodynamic force given by projections in the associated coordinate system; $\overline{M} = (M_x, M_y, M_z)$ — aerodynamic moment. Unknown reaction forces $\overline{N}^A, \overline{N}^B$ will be determined in the earth's coordinate system.

The equilibrium conditions of the airship are expressed in vector form as follow:

$$\begin{aligned} &\overline{N}^A + \overline{N}^B + \overline{G} + \overline{A} + \overline{R} = 0, \\ &\overline{Oa} \times \overline{N}^A + \overline{Ob} \times \overline{N}^B + \overline{OO_1} \times \overline{G} + \overline{M} = 0. \end{aligned} \tag{1}$$

The equation of moments is written with respect to the point O.

The scalar form of writing equations (1) makes it possible to obtain six conditions connecting the following twelve unknown quantities: x_g, y_g, z_g — parameters that determine the position of the pole of the airship in a fixed coordinate system; ϑ, ψ, γ— parameters that determine its angular position; $N^A_{x_g}, N^A_{y_g}, N^A_{z_g}, N^B_{x_g}, N^B_{y_g}, N^B_{z_g}$ — projections of reaction forces on the axes of the earth's coordinate system. The missing six equations can be obtained if we take into account the action of connections that restrict the motion of the device. If neither of the two bonds is released in the equilibrium position (all the cables are taut), the following two geometric relations must be fulfilled:

$$\overline{O_0A} = \overline{O_0O} + \overline{Oa} + \overline{aA},$$
$$\overline{O_0B} = \overline{O_0O} + \overline{Ob} + \overline{bB}. \tag{2}$$

Ignoring the slackness of the cables in the equilibrium position, i.e. assuming that their configuration in this position differs a little from a straight line and, consequently, the direction of the reaction forces coincides with the direction of the vectors and, it is obtained:

$$\overline{aA} = \frac{\overline{N}^A}{\left|\overline{N}^A\right|}\left(1 + k\left|\overline{N}^A\right|\right)l_A,$$
$$\overline{bB} = \frac{\overline{N}^B}{\left|\overline{N}^B\right|}\left(1 + k\left|\overline{N}^B\right|\right)l_B, \tag{3}$$

where, k — the coefficient of linear cable tension.

To loopback the resulting system of equations (1–3) it is necessary to define dependencies $\overline{R}(\alpha,\beta)$, $\overline{M}(\alpha,\beta)$, to express the angle of attack α and sideslip angle β using the desired unknowns and note that the vector quantities appearing in (1, 2) are defined in different coordinate systems.

Let's design the vectors $\overline{R}, \overline{M}, \overline{Oa}, \overline{Ob}, \overline{OO_1}$, defined in the associated coordinate system on the axis of the earth's coordinate system using the corresponding transformation matrix. Then

$$\begin{pmatrix} R_{x_g} \\ R_{y_g} \\ R_{z_g} \end{pmatrix} = Q\begin{pmatrix} R_x \\ R_y \\ R_z \end{pmatrix}, \quad \begin{pmatrix} M_{x_g} \\ M_{y_g} \\ M_{z_g} \end{pmatrix} = Q\begin{pmatrix} M_x \\ M_y \\ M_z \end{pmatrix}, \tag{4}$$

where $R_x = -c_x qU^{\frac{2}{3}}$, $R_y = c_y qU^{\frac{2}{3}}$, $R_z = c_z qU^{\frac{2}{3}}$, $M_x = m_x qU$, $M_y = m_y qU$, $M_z = m_z qU$,

$$c_x = c_{x0} + c_x^{\alpha^2}\alpha^2 + c_x^{\beta^2}\beta^2,\ c_y = c_{y0} + c_y^{\alpha}\alpha + c_y^{\alpha^2}\alpha|\alpha|,\ c_z = c_z^{\beta}\beta + c_z^{\beta^2}\beta|\beta|, \tag{5}$$

$$m_x = m_x^{\beta}\beta + m_x^{\alpha\beta}\alpha\beta,\ m_y = m_y^{\beta}\beta + m_y^{\beta^2}\beta|\beta|,\ m_z = m_{z0} + m_z^{\alpha}\alpha + m_z^{\alpha^2}\alpha|\alpha|,\ q = \frac{\rho W^2}{2},$$

W — modulus of the wind speed vector; U — the volume of the device.

Nonlinear terms in the expressions of aerodynamic coefficients can be taken in a different form, for example, in the form of a cubic dependence from α, β, which is not of fundamental importance for constructing the final mathematical model.

Wind speed vector can be written as follow;

$$\begin{pmatrix} W_x \\ W_y \\ W_z \end{pmatrix} = Q^T\begin{pmatrix} W_{x_g} \\ W_{y_g} \\ W_{z_g} \end{pmatrix}, \tag{6}$$

where $W_{x_g} = 0, W_{y_g} = 0$ since the problem is flat due to the above assumption.

For the angles of attack α and sideslip angle β, let's write the following equations

$$\alpha = -arctg\left(\frac{W_y}{W_x}\right),\ \beta = \arcsin\left(\frac{W_z}{W}\right). \tag{7}$$

When designing equations (1) on the axis of the earth's coordinate system, it is necessary to additionally take into account the following relations:

$$\overline{Oa} = Q\begin{pmatrix} x_a \\ y_a \\ z_a \end{pmatrix},\ \overline{Ob} = Q\begin{pmatrix} x_b \\ y_b \\ z_b \end{pmatrix},\ \overline{OO_1} = Q\begin{pmatrix} x_c \\ y_c \\ z_c \end{pmatrix}, \tag{8}$$

where (x_a, y_a, z_a), (x_b, y_b, z_b), (x_c, y_c, z_c) are elements of vectors $\overline{Oa}$, $\overline{Ob}$, $\overline{OO_1}$ in associated axes.

The system of nonlinear equations (1–9) is the desired mathematical model of the equilibrium of the airship in the parking lot. This is a closed system of equations with respect to the twelve unknowns:

$$x_g, y_g, z_g, \vartheta, \psi, \gamma, N^A_{x_g}, N^A_{y_g}, N^A_{z_g}, N^B_{x_g}, N^B_{y_g}, N^B_{z_g}.$$

The vector $\overline{W}$ is considered as known here.

4. Conclusion

The constructed nonlinear mathematical model is quite difficult to use, since it requires knowledge of the equilibrium position, which allows to correctly set the lengths of both cables. This follows from the fact that when setting the problem, the equilibrium position was supposed to be found among all possible positions of the system, at which both cables are taut. In real conditions, this requirement can be achieved by adjusting the length of the cables, and these adjustments will depend on the size and direction of the wind speed vector $\overline{W}$.

References

[1] Khoury G A and Gillett J D 1999 *Airship Technology* (Cambridge: Cambridge University Press)
[2] Khoury G A 1998 Review of unconventional design *24th Joint Annual Symposium of the Airship Association, June 26–28, Bedford* p 21
[3] Nayler A 1998 Airship activity and development world-wide *24th Joint Annual Symposium of the Airship Association* p 1
[4] De Laurier J D 1972 *J. Aircraft* **9** 646–51
[5] De Laurier J D, Lowe J D, Uffen D R and Amann J H 1986 Research on airship flight dynamics at the university of Toronto *Conf. on "Airship design and operation – present and future"* (Royal Aeronautical Society)
[6] Fichtl G H Problems in simulation of atmospheric boundari layer flows *AGARD CP-140*
[7] Houbolt J C 1973 Survey on effect of surface wind on aircraft design and operation and recomendations for needed wind research *NASA CR-2360*
[8] Ramos J, Bueno S, Maeta S, de Paiva E, Asamuma K, Nascimiento L, Bergerman M, Elfes A, Antonio J and Beiral R 1998 Project AURORA. *The 24th Joint Annual Symposium organized by the Airship Association, 26–28 June 1998, Bedford, England* p 91
[9] Oswald W B 1932 *The Transverse Force Distribution on Ellipsoidal and Nearly Ellipsoidal Bodies Moving in an Arbitrary Potential Flow. Thesis* (Calif. Institute of Technology)
[10] Bicho E, Moreira A, Carvalheira M and Erlhagen W 2005 *Autonomus Flight Trajectory Generation via Attractor Dynamics International Conference on Intelligent Robots and Systems IEEE/RSJ* p 615
[11] Bestaoui S Y 2012 Lighter than Air Robots *Intelligent Systems, Control and Automation* (Springer Science+Business Media BV)

[12] Knoblock F D and Troller Th 1935 *Tests on Effect of Sidewind on the Ground-Hanling of Airships* (Daniel Guggenheim Airship Institute)
[13] Ashford R L 1986 Military & civilian applications for tethered aerostats *Conference on Sounding Rockets, Ballons & related Space Systems, AIAA 7th, Oct, Ocean City Md, Collection of Technical Papers, New York NY* p 28
[14] Thompson N 1972 *Quart. J. R. Met. Soc.* **98** 745–62
[15] Rice C L, Korn A O, McPhetres G H and Hunsinger F V Some unique applications for tethered balloon systems *AIAA* **81**
[16] De Laurier J D 1977 *Prediction of Tethered-Aerostat Response to Atmospheric Turbulence. Engineering Notes* (Amsterdam, Netherlands)
[17] Jones S P and Krausman J A 1982 *Journal of Aircraft* **19** 679–86
[18] Jones S P 1989 Prediction of aerostat & airship mooring mast loads by nonlinear dynamic simulation *AIAA Lighter-Than-Air Systems Technology Conference, 8th, Jacksonville, Fl, Oct. 5–7* Technical Papers, Washington, DC, AIAA
[19] Grumondz V and Matus V 1998 Dynamics of an airship connected to earth with a rope *The 24th Joint Annual Symposium organized by the Airship Association, 26–28 June, Bedford, England* p 73
[20] Zasolov R 1985 *TsAGI* **2268** 31–46
[21] Arie M 1986 *Airships* (Kiev: Naukova dumka)
[22] Boiko Y 2001 *Aeronautics* (Moscow: MGUP)
[23] Kirilin A 2013 *Airsips* (Moscow: MAI-Print)
[24] Grumondz V ans Rudnev V 1989 Airship established horizontal circulation *22-nd Tsiolkovsky readings* (Moscow, IIET of the the USSR Academy of science) pp 103–7
[25] Grumondz V and Matus V 1999 *MAI Vestnik* **2** 60–6
[26] Loytsansky L 1970 *Liquid and gas mechanics* (Moscow: Nauka)
[27] Grumondz V 2019 Airship Balancing and Stability at Longitudinal Established Motion. *Workshop on Materials and Engineering in Aeronautics. IOP Publishing. IOP Conf.Series: Materials Science and Engineering* **476** 012013

IOP Conf. Series: Materials Science and Engineering 1027 (2021) 012008 doi:10.1088/1757-899X/1027/1/012008

Hovering control algorithm validation for a mobile platform using an experimental test bench

S Jatsun, O Emelyanova, B Lushnikov, A S Martinez Leon, L M Mosquera Morocho, A Pechurin and C A Nolivos Sarmiento

Southwest State University, Kursk 305040, Russia

asml1992@yandex.ru

Abstract. This paper deals with the validation of a hovering control algorithm for a mobile aerial platform executed on a quadcopter pattern. In order to test and evaluate the proposed hovering algorithm in a safe and controllable environment conditions, an experimental test bench has been introduced. As part of the hovering control algorithm implementation process, a brief description of the dynamics and control laws of the quadcopter is presented. The considered control technique in this paper is based on PID control method. A program code has been written using Arduino IDE based on Arduino UNO microcontroller. A description of the conducted indoor experiment for the validation of the hovering algorithm, and also the main results of the performed practical experiment are presented

1. Introduction

Currently the analysis of existing models of autonomous multi-rotor highly manoeuvrable small unmanned aerial vehicles (UAVs) weighing up to 5 kg is the most promising and demanded field of research in the military and civil spheres. A modern UAV is a flying laboratory, which includes an aerial vehicle and a hardware-software complex that allows operating the device in manual, semiautomatic or automatic mode, as well as receiving, storing and displaying the telemetry data received from the UAV. Small-sized UAVs are now being widely developed for different applications and operations such as search, inspection and sensing of remote areas, hazardous material recovery, real-time forest fire monitoring, disaster relief support, surveillance of sensitive areas (borders, ports, oil pipelines), etc. [1, 5, 6, 9].

Quadcopters have demonstrated a high manoeuvrability, efficiency and simplicity over the other exiting configurations of UAVs when performing different tasks. In terms of control, a quadcopter is an under-actuated system with six degrees of freedom but only four control inputs which are related to the rotor speeds [1, 2, 4, 7]. Thus, by adjusting the rotor speeds separately becomes possible to generate the required thrust and torques for driving the quadcopter in the space. To achieve propulsion in a particular direction, the axis of quadcopter must be tilted with respect to the vertical axis. The translational motion of the quadcopter is hence coupled with its angular orientation, making quadcopter dynamics and control complex [2, 8, 11].

Generally, the default position of a quadcopter is in hover mode, where the deviations from orientation and hover are calculated using an Inertial Measurement Unit (IMU) and other measuring systems respectively. Then this information is processed and filtered for feeding back into the system for an accurate control of the UAV [3, 8, 10].

This paper is devoted to the hovering control algorithm validation of an UAV type quadcopter based on PID control method. The quadcopter control algorithm has been implemented using an Arduino architecture. For our study a self-designed quadcopter prototype has been developed, which is equipped with Arduino hardware. The experimental part of our study aims with the optimization of the UAV type quadcopter stabilization near desired hovering position in the space by the use of an experimental test bench (Figure 5). The proposed test bench is indispensable to perform tests and evaluate the quadcopter hovering control algorithm minimizing commonly risks in a safe and controllable environment conditions [5].

This paper is organized as follows: Section 2 presents a brief description mobile platform dynamic; Section 3 is related to the hover control system design; Section 4 shows the practical implementation of the proposed quadcopter hover algorithm in an experimental test bench; Section 5 presents the simulation results obtained in this work; Section 6 concludes this purpose, such as some notes about future work.

2. Quadcopter dynamics

The scheme of the quadcopter is provided in Figure 1. Consider the motion of the convertiplane in a fixed cartesian coordinate system *OXYZ*. Let us assume that $O_1X_1Y_1Z_1$ is the local coordinate system, passing through the center of mass C of the body of the quadcopter and the centers of mass A_i of *i-th* electric drive with propellers, hereinafter referred to as the rotor. The orientation of the body in the air is set by the Euler Angles (roll φ, pitch *θ* and yaw *ψ* angles).

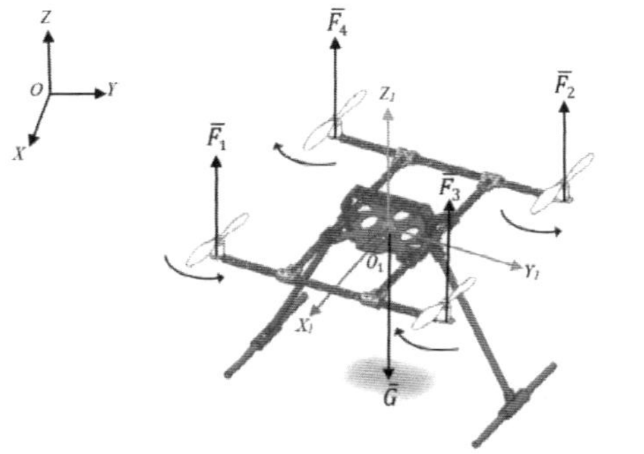

Figure 1. Calculation scheme of the UAV type quadcopter.

The forces applied to the quadcopter can be divided into forces that are set in the global coordinate system *OXY*: gravity force $\bar{G}$, applied to the center of mass *C*; and forces, defined in the local coordinate system CX_1Y_1: the forces generated by the propellers $\bar{F}_i$ (*I* = 1–4), applied at the center of mass of the rotors $\bar{A}_i$;

As result of the applied forces $\bar{F}_i$ to the rigid body by the motors the body of the quadcopter produces roll (φ), pitch (θ), and yaw (ψ) angles. According to the forces acting on the quadcopter rigid body, four control inputs can be defined [1–3]:

$$\begin{aligned} U_1 &= F_1 + F_2 + F_3 + F_4 \\ U_2 &= (-F_1 + F_2 + F_3 - F_4) \cdot L \\ U_3 &= (-F_1 + F_2 - F_3 + F_4) \cdot L \\ U_4 &= M_1 + M_2 - M_3 - M_4 \end{aligned} \quad (1)$$

where, F_i — Lift force generated by the *i* rotors, M_i — Torque produced by each rotor of the system, L — length of the arm from each rotor to the center of mass of the quadcopter, U_1 — lift force control input, U_2, U_3, U_4 — Euler angles φ, θ, *ψ* control inputs respectively.

From works [1–3, 7, 9] the govern dynamics of the quadcopter can be expressed in terms of the following system of equations, which governs the dynamics of the quagoverning dynamics of the quadrotor as:

$$
\begin{gathered}
\ddot{X}_1 = \frac{U_1}{m}(cos\varphi \cdot sin\theta \cdot cos\psi + sin\varphi \cdot sin\psi) \\
\ddot{Y}_1 = \frac{U_1}{m}(cos\varphi \cdot sin\theta \cdot sin\psi - sin\varphi \cdot cos\psi) \\
\ddot{Z}_1 = \frac{U_1}{m}(cos\varphi \cdot cos\theta) - g \\
\ddot{\varphi} = \frac{U_2}{I_{xx}} + \dot{\theta}\dot{\psi}\left(\frac{I_{yy}-I_{zz}}{I_{xx}}\right) - \frac{J_R}{I_{xx}}\dot{\theta}\dot{\Omega}_R \\
\ddot{\theta} = \frac{U_3}{I_{yy}} + \dot{\varphi}\dot{\psi}\left(\frac{I_{zz}-I_{xx}}{I_{yy}}\right) - \frac{J_R}{I_{yy}}\dot{\varphi}\dot{\Omega}_R \\
\ddot{\psi} = \frac{U_4}{I_{zz}} + \dot{\varphi}\dot{\theta}\left(\frac{I_{xx}-I_{yy}}{I_{zz}}\right)
\end{gathered} \tag{2}
$$

where, $X_I Y_I Z_I$ — coordinates, which describe the position of the center of mass of the quadcopter in the local system; φ, θ, ψ — Euler angles; m — mass of the system; I_{xx}, I_{yy}, I_{zz} — moments of inertia along the axes $X_I Y_I Z_I$ respectively, J_R — moment of inertia generated by the propellers of the rotors; Ω_R — angular velocity produced by the rotors and g — gravity constant.

In general conditions the system (2) is nonlinear and its solution is convenient to perform it using numerical methods. In our case, we will use this system of equations in the next section as part of the design of the quadcopter hover control algorithm in order to control the position of the UAV in the space.

3. Hover control system design

With the aim to achieve the hovering control of the quadcopte, the orientation and the altitude of the system must be tracked and controlled. Different control techniques have been implemented in the before mentioned works as we cited in the introduction of this work. Anyways, for practical implementation, performing a robust PID contoller is a suitable alternative. The results carried out with the use of a PID algorithm are competitive over other advanced methods and satiysfy the requirements. In Figure 2 a diagram block of the proposed control strategy is presented.

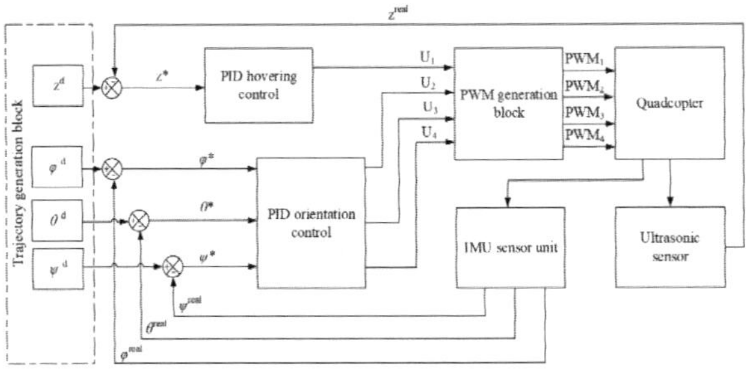

Figure 2. Quadcopter hovering control block diagram.

As we can see from Figure 2 the outputs of the system are the PWM signals, which make possible to control the speed of the quadcopter rotors. By controlling the rotor speed parameters, the quadcopter generates the required lift, orientation and altitude values for performing our task.

Besides, the presented hovering control block diagram compares the Euler angles and altitude desired values with the measured (real) values recived from the trajectory generation block, in this way becomes possible to calculate the respective errors. Thus, the measured error values are fed into the respective PID controller block to generate the needed hovering trust, and also the rolling, pitching and yawing control signals. According to the information presented in works [2, 10–12], the correction of the before mentioned signals should be done as follows:

$$
\begin{aligned}
U_1 &= K_p^z(z^{ideal} - z^{real}) + K_d^z(\dot{z}^{ideal} - \dot{z}^{real}) + K_i^z \int(z^{ideal} - z^{real})dt \\
U_2 &= K_p^\varphi(\varphi^{ideal} - \varphi^{real}) + K_d^\varphi(\dot{\varphi}^{ideal} - \dot{\varphi}^{real}) + K_i^\varphi \int(\varphi^{ideal} - \varphi^{real})dt \\
U_3 &= K_p^\theta(\theta^{ideal} - \theta^{real}) + K_d^\theta(\dot{\theta}^{ideal} - \dot{\theta}^{real}) + K_i^\theta \int(\theta^{ideal} - \theta^{real})dt \\
U_4 &= K_p^\psi(\psi^{ideal} - \psi^{real}) + K_d^\psi(\dot{\psi}^{ideal} - \dot{\psi}^{real}) + K_i^\psi \int(\psi^{ideal} - \psi^{real})dt
\end{aligned} \tag{3}
$$

where, K_p, K_i, K_d – PID controller coefficients, required for a controllable signal response; U_1, U_2, U_3, U_4 — lift force and Euler angles control inputs respectively; z^{ideal}, z^{real} — desired and real altitude values; $\varphi^{ideal}, \theta^{ideal}, \psi^{ideal}$ — commanded Euler angles values; $\varphi^{real}, \theta^{real}, \psi^{real}$ — measured Euler angles values.

Then, it is necessary to measure the altitude and orientation of the quadcopter in the space. In this work we are using two sensor units — an inertial sensor unit (IMU-650) and a ultrasonic distance sensor (HC-SR04).

As a suitable method for filtering the obtained data from the sensors is related to the use of a complementary filter [8]. A complementary filter passes the accelerometer signals through a low-pass filter and the gyroscope signals integral through a high-pass filter. Then, the obtained signals are added to estimate the orientation of the UAV in dynamic and static condition cases. The complementary filter block diagram is presented in Figure 3.

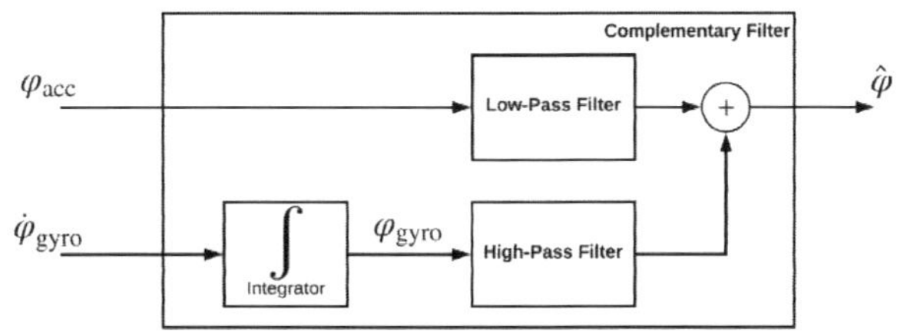

Figure 3. Complementary filter block diagram.

The code has been developed using Arduino architecture. In this case, we have used an Arduino UNO microcontroller. Also, a user interface for the visualization and validation of the results has been created. The main results of the implementation of the hovering control algorithm are presented in the section 5.

4. Quadcopter indoor experiment description

In this section, a brief description of the conducted indoor experiment is presented. With the aim to perform the experiment in a safe environment, the test has been completed in an experimental test bench. (Figure 5).

The quadcopter we are using for this experiment is a self-developed prototype created in the laboratory of the Department of mechanics, mechatronics and robotics at Southwest State University (Figure 4). The airframe of the prototype is completed in a symmetrical X configuration, its mass is equal to 1200 kg, the length of the arms (from the center of mass of the UAV to the center of mass of the rotors) is equivalent to 0.55 m. The lift force generated by each of the motors corresponds to 0.8 g. The autonomy of the prototype is approximately 35 minutes.

The indoor experiments were conducted in an experimental test bench. Thus, the quadcopter is fixed in the space, and the first tests can be carried out in a safe environment. The experiment is divided in 3 stages: I — take off, II — trajectory, III — hovering, IV — take-off. According to the dimensions and geometrical characteristics of the test bench, the maximum distance traveled by the quadcopter is around 2 meters during the indoor test. The prototype performs the experiment in a cycle of 60 seconds. During the validation process the cycle is carried out two times.

IOP Conf. Series: Materials Science and Engineering 1027 (2021) 012008 doi:10.1088/1757-899X/1027/1/012008

Figure 4. Quadcopter prototype.

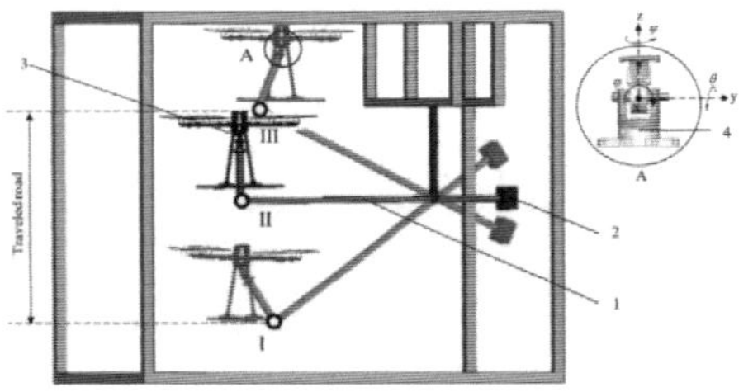

Figure 5. Experimental test bench:
1 — Support, 2 — Load, 3 — Prototype,
4 — Universal joint (3DOF).

During the realization of the experiment, the following assumptions have been considered:

- The real start position of the quadcopter is 0,85 m and the real end position is 1.75 m relative to the ground;
- The rotation around z axis (ψ angle) is set to zero degrees;
- The maximum roll (φ) and pitch (θ) angles range is limited to +/-30 degrees;
- The desired orientation of the quadcopter in the space relative to the local coordinate system during the experiment is zero degrees;
- External disturbances have been excluded.

5. Validation of the experimental results

In order to implement the proposed hovering control algorithm for the quadcopter. A simulation of the model has been carried out in Simulink/MATLAB [12]. The simulation was performed taking into account the mathematical model of the quadcopter and the real parameters of the electromechanical components of the system. This study provided a better understanding of the quadcopter control. Also, as part of the simulation results, we obtained the information about the coefficients of the controller for hovering task. Hence, during the practical implementation of the hovering algorithm, the simulation-obtained parameters conducted to an unstable hovering of the quadcopter. This problem could be caused by the exclusions token into account during the mathematical modelling process. Thus, during our experiment, the parameters of the controller were improved empirically, based on theoretical control effects [4, 6, 11].

The experiment was carried out in an indoor environment using the setup depicted below to enhance safety and the accuracy of the measurements (Figure 6).

(a) (b)

Figure 6. Indoor experiment setup: (a) — front view, (b) — lateral view.

The main results obtained from the performed practical experiment are presented in Figures 7–10.

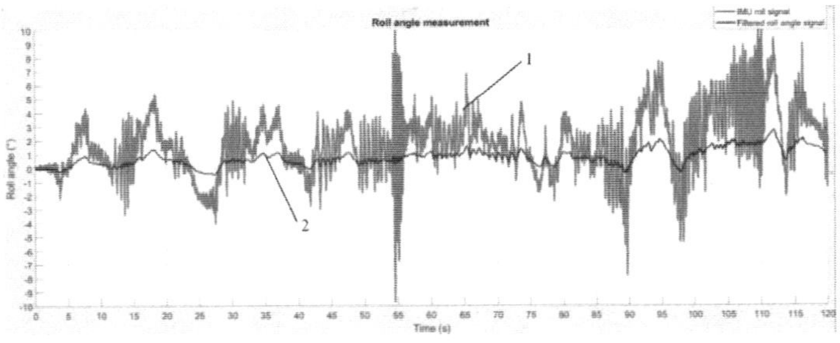

Figure 7. Roll angle measurement:
1 — IMU roll raw signal,
2 — Filtered roll angle signal.

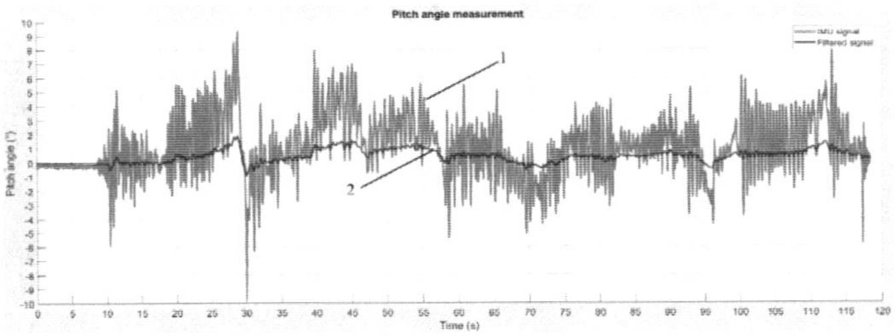

Figure 8. Pitch angle measurement:
1 — IMU pitch raw signal,
2 — Filtered pitch angle signal.

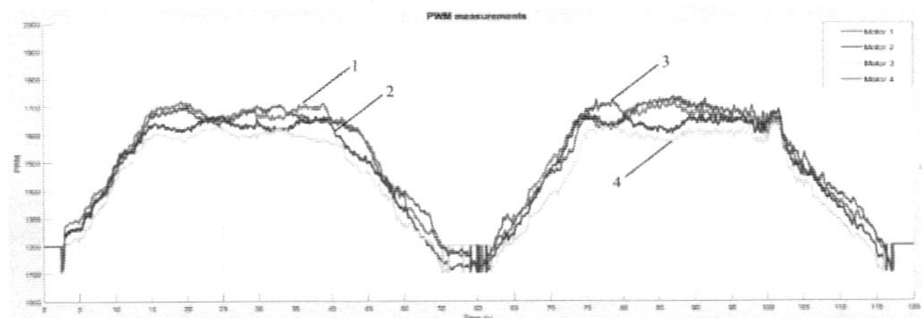

Figure 9. PWM generation signals for the motors M_1, M_2, M_3, M_4 respectively.

Figure 10. Quadcopter hovering control: 1 — Altitude measured by the ultrasonic sensor,
2 — Desired hovering trajectory.

Once the results of the indoor experiment of the UAV type quadcopter have been carried out, we can conclude that developed system presents a very sensitive characteristic to external factors. In order to achieve the stabilization of the quadcopter it is necessary to perform a high precise configuration of the PID regulators of the system. An inaccurate configuration of the regulators can affect significantly to the stability of the quadcopter (roll over or crash) [12].

6. Conclusions

In this paper the validation of a proposed hovering control algorithm for a mobile aerial platform executed on a quadcopter pattern has been performed. The introduced experimental test bench was used as an instrument to enhance safety conditions during the performance of the indoor experiment. The simulation-obtained parameters of the model from Simulink/MATLAB conducted to an unstable hovering of the quadcopter, the parameters of the controller were improved empirically, based on theoretical control effects. The obtained practical results satisfy the main requirements of the described task. Further, there will be implemented an adaptive method of control to adjust the PID parameters values, for example by using a neural network. With the aim to obtain a completely autonomous controller, an outdoor test is planned to be performed. The results of this test will be presented in further publications.

References

[1] Emelyanova O V et al. The synthesis of electric drives characteristics of the UAV of "convertiplane-tricopter" type *MATEC Web of Conferences* **99** 249–263 (2017)

[2] Mellinger D, Kumar V Control and Planning for Vehicles with Uncertainty in Dynamics. In: *Proceedings of the IEEE International Conference on Robotics and Automation (ICRA)*, 960–965 (2010)

[3] Jatsun S F et al. Investigation of Oscillations of a Quadcopter Convertiplane in Transient Mode in the Vertical Longitudinal Plane. In *Proceedings of 14th International Conference on Electromechanics and Robotics "Zavalishin's Readings"*, Springer, Singapore 345–358 (2020)

[4] Jatsun S et al. 2017 Control flight of a UAV type tricopter with fuzzy logic controller In *XIII International scientific and technical conference "Dynamics of Systems, Mechanisms and Machines", IEEE* 1–5 (2017)

[5] Jatsun S F, Emelyanova O V, Martinez Leon A S Design of an Experimental Test Bench for a UAV Type Convertiplane *IOP Conference Series: Materials Science and Engineering* **714 (1)** 012009 (2020)

[6] Sa I and Corke P 2012 System Identification, Estimation and Control for a Cost Effective Open-Source Quadcopter *Internal Conference on Robotics and Automation, IEEE* Minnesota 2202–2209 (2012)

[7] Valeria E, Caldera R., Lara S and Guichard J LQR Control for a Quadrotor using Unit Quarternions: Modeling and Simulation, IEEE, Puebla, Mexico (2012)

[8] Martinez Leon A S, Jatsun S F and Emelyanova O V Control of the electric drives of a multirotor system type convertertiplane *Fundamental and Applied Problems of Technics and technology Journal* **1** 83–93 (2020)

[9] Jatsun S et al. Modeling and control architecture of an autonomous mobile aerial platform for environmental monitoring In *2019 International Conference on Information Systems and Computer Science (INCISCOS)*, IEEE, Quito, Ecuador 177–182 (2019)

[10] Suresh H, Sulficar A and Desai V Hovering control of a quadcopter using linear and nonlinear techniques *International Journal of Mechatronics and Automation* **6** 120–129 (2018)

[11] Doukhi O, Fayjie A R and Lee D J 2017 Intelligent controller design for quad-rotor stabilization in presence of parameter variations *Journal of Advanced Transportation* **2017**

[12] Jatsun S et al. Synthesis of SimMechanics Model of Quadcopter Using SolidWorks CAD Translator Function *Proceedings of 15th International Conference on Electromechanics and Robotics" Zavalishin's Readings"*. Springer, Singapore. pp 125–137

Nonlinear stochastic estimation of altitude and vertical flight speed in real time

V P Kharkov[1], O N Korsun[2,3], M N Pravidlo[3] and Moung Htang Om[3]

[1] Experimental Workshop NaukaSoft, 9 Godovikova, Moscow 129085, Russia
[2] State Research Institute of Aviation Systems, 7 Viktorenko, Moscow 125319, Russia
[3] Moscow aviation institute (National Research University), 4 Volokolamskoe Shosse, Moscow 125993, Russia

charkovvp@rambler.ru

Abstract. The optimal algorithm for calculating the altitude and vertical flight speed of an aircraft, which is based on complex processing of information from static pressure sensors, outdoor air temperature, and vertical acceleration is described. Information processing is performed using nonlinear estimation methods. Adaptive properties of the estimation algorithm are achieved by including into the state vector the parameters that characterize the properties of sensors and static pressure transmission paths.

1. Introduction

Various approaches are known to determine the altitude and speed parameters of an aircraft, and above all, the altitude and vertical flight speed [1, 2]. Based on the kinematic relations connecting height and vertical acceleration [3], the main relations of linear filters are obtained [4, 5], and complex processing for the identification of aerodynamic coefficients is described in [6].

However, none of these approaches solves the problem of accurately determining the altitude and vertical flight speed because they do not take into account the properties of the sensors and the pressure transmission paths. Implementation of these features into estimation algorithm leads to the problem of nonlinear estimation [6, 7]. The present paper deals with the synthesis of an optimal adaptive algorithm for processing signals from static pressure, air temperature, and vertical accelerometer sensors based on nonlinear estimation methods. Adaptive properties of the algorithm are achieved by including into the state vector of the estimation system the parameters that characterize the properties of sensors and static pressure transmission paths.

2. Problem statement

It is known that the flight altitude of the aircraft is determined by the measured values of static pressure, which is converted to the altitude value [8]. The receiver and static pressure sensor are connected by a pressure transmission air duct of finite known length. The static pressure at the output of the air duct is related to the actual air pressure by the transfer function of the form

$$W(P)=\frac{1}{\tau_p S+1}e^{-\beta S}, \tag{1}$$

where, τ_p — the time constant of the air duct, β — the time shift equal to the ratio of the length of the air duct to the speed of sound, S — the variable of the Laplace operator.

In general, the static pressure at the output of the converter is determined by the expression $P_p = P_\partial + \varpi_p + \Delta P$. Here ϖ_p — a random process of the "white" noise type with the intensity Q_p, P_p — the measured static pressure at the current height, P_∂ — the static pressure at the outlet of the air duct determined by the solution (1), ΔP — a slowly changing random error.

Assuming that air is an ideal gas, the kinematic relations between vertical acceleration and static pressure P are determined by the following equations:

$$\frac{dP}{dt} = -\frac{gP}{RT}V_y, \quad \frac{dV_y}{dt} = a_y, \tag{2}$$

where, R — the gas constant, g — the acceleration due to gravity, a_y — the vertical acceleration, T — the outside air temperature at a given altitude.

According to the ideal standard atmosphere model the outdoor air temperature has a rather complex dependence on the flight altitude. For example, in the surface layer of the atmosphere up to altitude of H=11000 m the function T=f(H) may be represented by a typical form:

$$\text{T}=T_0 + \beta_T(H - H_0),$$

where, H_0, T_0 — the altitude and temperature corresponding to the bottom of the interval, that is to the zero altitude, β_T — the temperature gradient, the value of which for this altitude interval in an ideal atmosphere is –0.0065 K/m. In real conditions, this value may change in one direction or another. On board of an aircraft, the outdoor air temperature is measured through the temperature of the air flow, which is related to the actual temperature by the function $T_в = KT$, where K is the proportionality coefficient. Then for the aircraft, it can be written that the measured value of the temperature $T_в$ with the outdoor air temperature T is related by the equation:

$$\frac{dT}{dt} = \beta_T V_y, \quad \frac{dT_в}{dt} = \frac{KT - T_в}{\tau_T}, \tag{3}$$

where, τ_T — the time constant of the temperature sensor.

The values of vertical acceleration, static pressure, and outdoor air temperature at the current flight altitude are measured by appropriate sensors, which have both instrumental and methodological errors. Models of errors of the accelerometer, pressure and temperature sensors (1), (2) and (3) can be presented as,

$$\Delta a_y = a_1 + \xi_y; \; \Delta p = p_1 + \xi_p; \; \Delta T = T_1 + \xi_T, \tag{4}$$

where, a_1, p_1, T_1 — the slowly changing components of the corresponding errors, ξ_y, ξ_p, ξ_T — the stochastic components of the errors, which are random processes with correlation functions

$$K_y(\tau) = D_y e^{-\lambda_y|\tau|}, \; K_p(\tau) = D_p e^{-\lambda_p|\tau|}, \; K_T(\tau) = D_T e^{-\lambda_T|\tau|}. \tag{5}$$

Here, D_y, D_p, D_T — the variances, λ_y, λ_p, λ_T — the correlation radius of the corresponding random processes. Evidently, the assumption (5) means that the sensors stochastic errors are the «coloured» noises.

Set of equations (1), (2), and (5) defines the model of the system, the state vector of which must be estimated based on the samples of pressure, temperature, and acceleration sensors.

3. Basic mathematical relations

It is known that the time constants τ_p and τ_T depend on the altitude and speed of flight in accordance with the expressions approximating them:

$$\tau_P = \frac{K_P P_0}{P}, \ \tau_T = \frac{K_T P_0 \theta(M)}{P}, \tag{6}$$

where, P_0 — the static pressure at the altitude $H = 0$, K_P, K_T — the coefficients of the "sliding" approximation, $\theta(M)$ is a known function. The slowly changing error of the accelerometer is assumed to be constant with a zero mean and known variance, which will later be included in the state vector in order to clarify it. It should be noted here that the vertical acceleration itself will not enter the state vector and is further treated as a parameter or as an external known signal with its own distribution parameters.

To ensure the adaptive properties of the algorithm for calculating the altitude and speed of the flight, it is proposed to enter the parameters τ_p, τ_T, K_P and K_T into the state vector of the system. Then the generalized mathematical model of the estimation system, taking into account (1), (2) and (5), will take the form:

$$\begin{gathered}
\frac{dP}{dt} = -\frac{gP}{RT}V_y, \quad \frac{dV_y}{dt} = a_y + \Delta a_y, \quad \frac{dT}{dt} = \beta_T V_y, \quad \frac{dT_{\text{в}}}{dt} = \frac{KT - T_{\text{в}}}{\tau_T}, \\
\frac{d}{dt}P_p = \frac{P(t-\tau_p)t - P_p}{\tau_p} + \varpi_P, \quad \frac{dT}{dt} = \beta_T V_y, \quad \frac{dT_{\text{в}}}{dt} = \frac{KT - T_{\text{в}}}{\tau_T}, \\
\frac{d\tau_P}{dt} = -\frac{K_P g P_0 V_y}{RTP^2}, \quad \frac{d\tau_T}{dt} = -\frac{K_T g P_0 V_y \theta(M)}{RTP^2}, \\
\frac{dK_T}{dt} = 0, \quad \frac{dK_P}{dt} = 0, \quad \frac{d\beta_T}{dt} = 0 \quad \frac{da_1}{dt} = 0,
\end{gathered} \tag{7}$$

where, P_p — the value of the static pressure at the output of the air duct.

Temperature, pressure, and acceleration measurements are performed at discrete time points t_K with a constant step, the mathematical description of which depends on the method of accounting for random errors (5) in the measurement model. The simplest way to represent a measurement model is to approximate (5) with "white" noise with a given variance. Then:

$$Z_1(t_K) = P_p(t_K) + \xi_p(t_K), \ Z_2(t_K) = T_{\text{В}}(t_K) + \xi_p(t_K), \ a_y^* = a_y + a_1. \tag{8}$$

Random processes $\xi_P(t_K)$ and $\xi_T(t_K)$ are centered discrete "white" noises with known distributions:

$$M\left[\xi_P(t_K)\xi_P(t_{K+n})\right] = D_P\delta(k-n), \ M\left[\xi_T(t_K)\xi_T(t_{K+n})\right] = D_T\delta(k-n).$$

Here, $\delta(k-n)$ — the Kronecker symbol.

Let's write the equations (7) and the measurement system (8) in matrix form,

$$\begin{gathered}
\frac{dX}{dt} = f(X) + Ba_y^* + \varpi(t), \\
Z(t_K) = CX(t_K) + \vartheta(t_K).
\end{gathered} \tag{9}$$

The measurement vector $Z(t_K)$ includes two components: static pressure and outdoor air temperature.

The problem of determining the altitude and vertical flight speed, taking into account the representation (9), is formulated as follows. According to the observations $Z(t_K)$, $K=1, 2, 3, ..., n$ it is required to determine estimates of the state vector of system (9) $X(t_K)$, assuming that its initial state is normally distributed, $X(t_0)\in N(\bar{X}_0, P_0)$, where $(\bar{X}_0, P_0)$ — a priori values of mean and correlation matrix of the vector $X(t_0)$.

4. The adaptive algorithm of optimal estimation

To solve the nonlinear estimation problem we use a discrete-continuous extended Kalman filter [3], the main feature of which is that the state forecast is determined in accordance with the system equation, and the residual gain and error correlation matrices are calculated at discrete times.

Then the algorithm for determining estimates can be represented as the following sequence.

1. Calculating the forecast $\overline{X}$ of the state vector $X(t)$ for time interval $t \in [t_K, t_{K+1}]$

$$\frac{d\bar{X}(t)}{dt} = f(\bar{X}(t) + Ba_y(t_K), (t_K)\,, \overline{X} = \hat{X}_K\,.$$

2. Calculation of the correlation matrix of the forecast errors:

$$P(t_{K+1}, t_K) = \Phi(t_{K+1}, t_K)V_K\Phi(t_{K+1}, t_K)^T + W(t_{K+1}),$$

where

$$d\Phi(t,t_K)/_{dt} = A(t)d\Phi(t,t_K), \quad A(t) = \partial f(\bar{X}(t)/_{\partial\bar{X}(t)},$$

$$\Phi(t_K, t_K) = I\,, \; t \in [t_K, t_{K+1}]\;.$$

The correlation matrix $W(t_K)$, a discrete white noise that simulates the effect of a "standing" wave when transmitting pressure in the air duct, is determined by solving an equation for $t \in [t_K, t_{K+1}]$,:

$$\frac{dW(t)}{dt} = AW(t) + W^T(t)A + S_\varpi, \quad W(t_K) = 0.$$

The intensity S_ϖ of white noise is determined from the flight data.

3. Determining the gain

$$K_{K+1} = P(t_{K+1}, t_K)C^T[CP(t_{K+1}, t_K)C^T + R]^{-1}, \qquad R = \mathrm{diag}(D_P, D_T).$$

4. Calculation of the correlation matrix of the estimation error

$$V_{K+1} = [K_{K+1}C - I]P(t_{K+1}/t_K)[K_{K+1}C - I]^T + K_{K+1}RK_K + Q_{K+1}.$$

5. Determining the current estimate of the state vector

$$\hat{X}_{K+1} = \bar{X}_{K+1} + K_{K+1}(Z_{K+1} - C\bar{X}_{K+1}).$$

The value of the current flight altitude is determined by the static pressure estimate in accordance with the expression:

$$H(T) = \frac{T_0}{\beta_T}\left[\frac{P^{-\frac{R\beta_T}{g_0}}}{P_0} - 1\right],$$

where, P_0, T_0 — the standard air pressure and temperature at sea level, g_0 — the gravity acceleration.

The described algorithm works in the range of altitudes no more than 11000 m. For other ranges of altitudes, it is necessary to take into account the nature of changes in the temperature gradient.

5. Research result

The efficiency of the estimation algorithm was evaluated based on the results of numerical simulation for the following values of noise parameters: $D_p = 169$ Pa^2, $D_a = 0.04$ м^{2}c^4, $D_E = 9$ K^2. The time constants were assumed to be: $\tau_p = 1.44$ c, $\tau_E = 1.96$ c and the temperature gradient $\beta = 0.012$ $K м^{-1}$.

The dynamics of the processes of estimating the altitude and flight speed parameters of the aircraft is shown in figure 1. The figure shows the change in the errors of estimating the altitude ΔH, vertical velocity ΔV_y, and the standard deviation of the vertical velocity estimate at level 3σ.

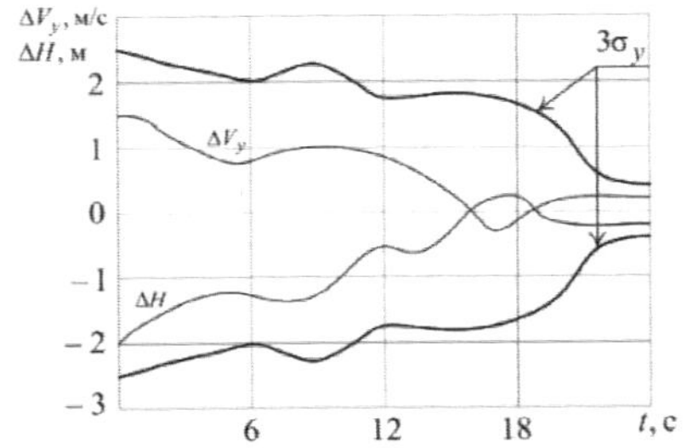

Figure 1. Graphs of changes in altitude and vertical flight speed estimation errors.

The flight altitude was determined by solving the first equation of the system (5) under the assumption that the value of g does not depend on the flight altitude, the outdoor air temperature changed according to $T = T_0 + \beta_H$, where $T_0 = 288.15$ K is the normal air temperature at zero altitude. Note that the constant error of static pressure measurement for a given structure of the measuring system cannot be estimated and, therefore, it is not included in the state vector of the estimation system (7).

Figure 2 shows the process of estimating the time constant of the air duct. Its dependence on altitude is also shown here. It should be noted that the convergence rate of the identification algorithm is proportional to the vertical flight speed.

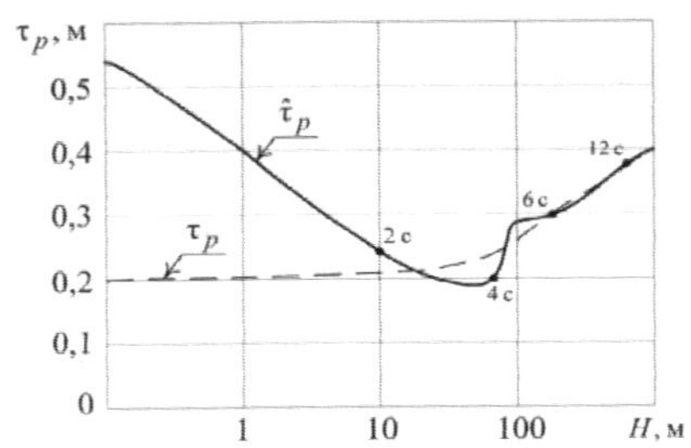

Figure 2. Graph of changes in the estimation of the time constant of the air duct.

6. Conclusion

In conclusion, it is noted that the proposed algorithm allows estimating the altitude and flight speed parameters of the aircraft while simultaneously identifying the internal parameters of the measurement system. The rate of convergence of estimation algorithms depends on the type of maneuver performed, and the more energetic the maneuver, the higher the rate of reducing a priori errors in estimated parameters.

Acknowledgements
This work was supported by Russian Foundation for Basic Research (RFBR), project 20-08-00449-a.

References

[1] Pushkov S G, Lovitskii L L and Korsun O N 2018 Aerodynamic errors of the systems aimed at measuring the static pressure of an aircraft in the sliding modes of flight *Measurement Techniques* **61** (2) 140–7

[2] Korsun O N, Tulekbayeva A K and Toktabek A A 2017 Estimation of errors of aircraft air parameters measurements based on satellite navigation system data *Proc. 2017 2nd Int. Ural Conf. on Measurements URALCON 2017* (IEEE, Inc.) p 145–8
[3] Chernodarov A V 2017 *Monitiring, diagnostics and identifications of avionic systems* (Moscow, Nauchtekhizdat) p 299
[4] Arkhipov A S and Semenikhin K V 2019 Confidence analysis of linear unbiased estimates under uncertain unimodal noise distributions *Journal of computer and systems sciences international* **58** (5) 674–83
[5] Zhirabok A N, Shumskii A E and Zuev A V 2019 Diagnosis of linear systems based on sliding mode observers *Journal of computer and systems sciences international* **58** (6) 898–914
[6] Korsun O N, Om M H, Latt K Z and Stulovskii A V 2019 Real-time aerodynamic parameter identification for the purpose of aircraft intelligent technical state monitoring *Procedia Computer Science 12th Int. Symp. Intelligent Systems, INTELS 2016* p 67–74
[7] Klein V and Morelli E A 2006 *Aircraft system identification: Theory and practice* (Reston, VA: AIAA, Inc.)
[8] *Aerodynamics, stability and controllability of supersonic aircraft* 1998 ed G S Bjushgens (Moscow: Nauka) p 16

Layered composite for a geopolygon mire

A P Kondratov, G N Zhuravleva and Y M Syltanova

Moscow Polytechnic University, 38 Bolshaya Semyonovskaya str., Moscow 107023, Russia

a.p.kondratov@mospolytech.ru

Abstract. We have investigated flexible composite materials of test objects (mires) intended for setting the aviation optical equipment for remote sensing of the Earth. The brightness fields and control images of the mire contrasting elements are obtained by the method of ink-jet printing on banner fabric. We have proposed a new layered structure of the material for the operating contrast control. We have established quantitative patterns of changes in the optical characteristics of test objects when using an outer layer made of a transparent polymer film, which makes it possible to vary the characteristics of test objects for adjusting thermal imaging devices and aviation optical equipment using low-cost polygraphic methods.

1. Introduction

One of the most important interdisciplinary areas of modern applied physics and materials science is the development of systems for multivariable integrated control of thermal imaging devices and the creation of test objects for setting the onboard aircraft equipment [1–3]. Full-scale tests of such systems are always associated with significant financial costs and interdepartmental organisational arrangements. As a result, the shortages of design solutions and errors in the choice of materials for test objects are revealed at the final stage of testing.

The object of this paper is to develop the structure of a layered composite with minimal material costs and the ability to vary the contrast of the radiation temperature of the geopolygon mire elements.

A test object know, a passive geopolygon mire for setting and determining the parameters of onboard aircraft equipment in the infrared range [4, 5], contains rectangular strips of thin aluminum tubes with a light-return surface, the strips have recesses in the form of corner reflectors adjoining each other with the edges of recesses [6].

Aluminum tubes with a knurled surface provide noise immunity of the test object from side illumination and control the radiation temperature of IR-mire strips.

The disadvantage of the learned passive geopolygon infrared mire is the use of significant amount of non-ferrous metal, a high probability of mechanical damage (deformation) of thin tubes, the impossibility of prompt installation and dismantling if it is necessary to change the direction of laying the stripes depending on the direction of aircraft movement.

In recent years, test objects made of polymer and composite materials that are easily and quickly erected on the earth's surface have been created and patented [7], these materials are obtained by high-performance methods of processing thermoplastic polymers into films, fabrics, and nonwoven fibrous materials. A layer of aluminum or paint (using printing equipment) is applied to flexible polymer fabrics.

Monochrome images of a certain optical density are printed in order to obtain the specified contrast values for the radiation temperature or the reflection coefficient of electromagnetic radiation. So, for example, a mire for setting and determining the parameters of optoelectronic systems [8–10] contains several dashed images of various sizes applied to the surface of a flexible material of contrasting color at different angles to the intended direction of flight. The device consists of a main rectangular fragment made of a fabric-reinforced polymer material of dark color, with width multiple of three meters with light images of groups of rectangular dashes applied to it, located at angles of 0°, 45°, 90° and 135° to the side of the fragment, and rectangular auxiliary unified fragments of the side background. The dashed elements of the test object are made of a nonwoven polymeric material with an identical spectral reflection coefficient of electromagnetic radiation to the background areas and containing regularly repeating recesses located at an equal distance from each other.

2. Subjects and methods

Material: banner fabric S-PRW334240B made in the Republic of Korea from a mesh of polyester threads coated with PVC plastisol.

The water-base ink was applied to a transparent film of polyethylene terephthalate (lavsan) made by Emtec using a Canon Pixma MG2540S ink-jet printer.

The optical density and reflection coefficient of D65 built-in light source at an observation angle of 2 degrees were determined using an X-Rite Pantone e-Xact spectrophotometer with GretagMacbeth KeyWizard V2.5 software. Banner fabric without a layer of paint was used as a white balance standard. We carried out 10 identical measurements over the entire area of the presented sample and calculated the mean value of the changeable values, the mean root square deviation of the optical density and the reflection coefficient which are shown in the diagram (Figure 3).

3. Results and discussion

Figure 1 shows a cross-section of a non-woven fibrous fabrics which is used for the production of the lower layer of the test object by the method of hot calendering of the fabric made of thermoplastic fibers, the regularly spaced prismatic recesses are formed on the lower layer forming the background of the mire. The fabric surface is covered with a mirror layer of aluminum, which provides an effective reflection of light and heat radiation. The shape, size of the recesses and the frequency of their location provide uniform radiation scattering. To obtain contrasting images, the stripes of polymer material painted in light colors are applied to the background fabric. To vary the contrast of the elements of the test object in [2], it was proposed to perforate the outer layer so that the integral flux of radiation reflected from the stripes would make up a certain fraction of the radiation flux reflected from the background or exceed it by a certain amount.

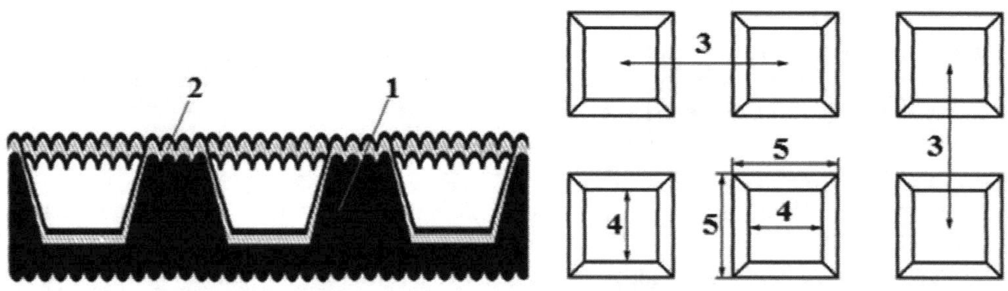

Figure 1. Macrostructure of mire elements made of metallized non-woven fibrous fabric with recesses. 1 — polypropylene fibers, 2 — aluminum layer; 3 — the distance between the recesses; 4, 5 — frontal dimensions of the recesses.

We propose to use translucent prints on a polymer film to create a certain optical density of dashes and regulate the contrast of the target, instead of perforations of dashed elements, the film should be applied to a background canvas with a certain optical density or radiation temperature.

Printing on polymer film according to the "roll to roll" technological scheme is a high-performance printing process and can be used to make an easily removable outer layer of the mire [5]. Prompt replacement of the outer layer of a transparent film with geometric shapes applied to it allows to change the contrast of the mire elements.

Figure 2 shows a frontal view of the dashed element of the mire — a film, printed by alternating circles or squares, the total area of which is 20 ÷ 40%.

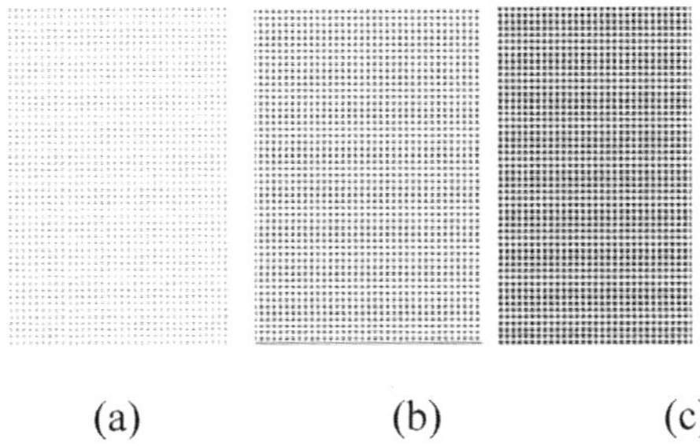

Figure 2. View of dashed elements of the mire with a proportion of the printed surface: (a) — 20%; (b) — 30%; (c) — 40%.

The prints were used to estimate the optical characteristics (optical density and light reflectance) of the samples. When determining the optical density of the printed film, a layer of banner fabric of white, gray or black color was placed under the film. The results of these measurements and calculations are shown in the table

To calculate the total proportion of the area of prints the following formula is used:

$$\alpha = [\rho_f (k+1) - \rho_{mp}] / (\rho_{mp} - \rho_f), \tag{1}$$

where k is the contrast of the printed stripes in terms of the radiation temperature or the reflection coefficient of electromagnetic radiation; ρ_f — integral reflection coefficient of the radiation flux from the background fabric; ρ_{mp} — integral coefficient of electromagnetic radiation reflection from the printed strip of the mire.

The contrast of the dashed mire is determined by the ratio:

$$K_M = (\rho_{mp} - \rho_f)/\rho_f, \tag{2}$$

where ρ_f is the reflection coefficient of the background (banner fabric).

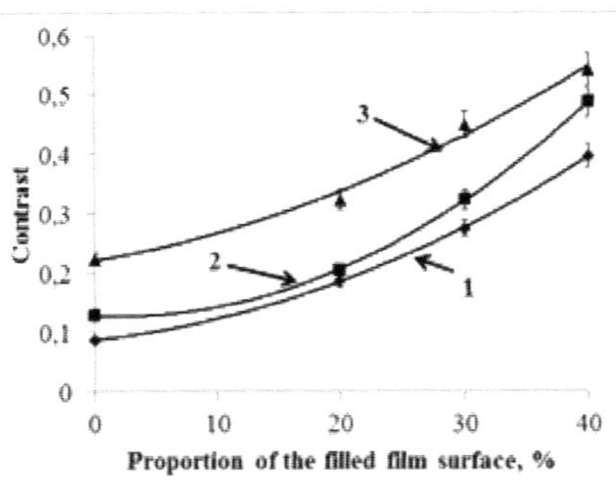

Figure 3. Dependence of the stripes contrast on the proportion of the filled (printed) film surface for white banner fabric: 1 — no air gap; 2 — air gap h = 0.14 mm; 3 — air gap h = 0.37 mm.

In the laboratory version, the optical density of the printed film and the reflection coefficient were estimated. Contrast was calculated based on the measurement results. We evaluated the possibility of the influence of the air gap between the white banner fabric (background) and the printed film (strip) on the optical characteristics of the material (Table 1).

Table 1. Optical characteristics of white banner fabric.

Material	For printed film surface, %	Gap	Optical density	Reflection coefficient	Contrast
White banner fabric			0.06	0.871	
White banner fabric + film	0		0.10	0.794	0.088
		h = 0.14 mm	0.12	0.759	0.129
		h = 0.37 mm	0.17	0.676	0.224
	20		0.15	0.708	0.187
		h = 0.14 mm	0.16	0.692	0.206
		h = 0.37 mm	0.23	0.589	0.324
	30		0.20	0.631	0.276
		h = 0.14 mm	0.23	0.589	0.324
		h = 0.37 mm	0.32	0.479	0.450
	40		0.28	0.525	0.397
		h = 0.14 mm	0.35	0.447	0.487
		h = 0.37 mm	0.40	0.398	0.543

4. Conclusion

When the air gap size in a multilayer package made of polymeric materials is increased, the optical characteristics for all three colors of the banner film are increased, too, as well as the proportion of the printed surface is increased in all cases. So, as the air gap increases, the surface becomes darker.

The application of a multilayer package of polymer materials with interlayer printing provides an increase in the level of physical and mechanical properties of the product, thereby extending the operating time, but reduces the possibility of adjusting the optoelectronic equipment of aircraft.

References

[1] Zhuravleva G N, Nagornova I V, Kondratov A P, Bablyuk E B, Varepo L G 2017 Environmental durability diagnostic for printed identification codes of polymer insulation for distribution pipelines *AIP Conference Proceedings* **1876** 020077

[2] Kondratov A P, Nikolaev A A, Kondratova M A, Savenkova I A 2019 The influence of solar radiation on the parameters of the polymer type test pattern, designed to adjust the optical aircraft equipment *IOP Conference Series Materials Science and Engineering* **476** 012014

[3] Altukhov E V, Bablyuk E B, Veselov U G, Vladychenko O V, Moroz V A, Pat. 175973, RF, Dismountable test object, i.e. type test pattern to determine the parameters of electrooptical systems IPC G01M 11/02, G03B 43/00, H04N 17/00, pubd. 25.12.2017, Report No. 36 dated 17.04.2017

[4] Kondratov A P, Varepo I G, Nagornova I V, Ermakova I N 2015 Transparent layered materials based on variable color polyolefins *Procedia Engineering* **113** 423–8

[5] Lee J, Shin K, Lee C 2015 Analysis of dynamic thermal characteristic of register of roll-to-roll multi-layer printing systems *Robotics and Computer-Integrated Manufacturing* **35** 77–83

[6] Rader Bowers L M, Schmidtke Sobeck S J 2016 Impact of medium and ambient environment on the photodegradation of carmine in solution and paints *Dyes and Pigments* **127** 18–24

[7] Pickett J E, Gibson D A, Gardner M M 2008 Effects of irradiation conditions on the weathering of engineering thermoplastics *Polymer Degradation and Stability* **93** 1597–606

[8] Gherardi R, Becerril R, Nerin C, Bosetti 2016 Development of a multilayer antimicrobial packaging material for tomato puree using an innovative technology *O LWT — Food Science and Technology* **72** 361–7

[9] Zolesi B et al 2008 A new campaign for oblique-incidence ionospheric sounding over Europe and its data application *Journal of Atmospheric and Solar-Terrestrial Physics* **70** 854–65

[10] Weixiu Z, Qixin Z, Haichang Z, Xiaoning Q I 2018 One-coat epoxy coating development for the improvement of UV stability by DPP pigments *Dyes and pigments* **151** 157–64

IOP Conf. Series: Materials Science and Engineering 1027 (2021) 012011 doi:10.1088/1757-899X/1027/1/012011

Detection of dynamic errors in aircraft flight data

O N Korsun[1,2], S V Nikolaev[2] and Moung Htang Om[2]

[1] State Research Institute of Aviation Systems, 7 Viktorenko, Moscow 125319, Russia
[2] Moscow Aviation Institute (National Research University), 4 Volokolamskoe Shosse, Moscow 125993, Russia

marmotto@rambler.ru

Abstract. The paper considers the algorithmic and methodologic support for detecting dynamic errors in aircraft on-board measurements using parameter identification and system approach to flight data analysis. Examples of practical application of the considered methods and algorithms for detecting dynamic errors in modern aircraft flight data are presented.

1. Introduction

High accuracy of on-board measurements of flight parameters and correct estimation of errors that occur plays the key role in flight test analysis and aircraft technical state monitoring. The on-board measurement and recording systems generally meet the requirements for measurement accuracy, but in some cases special approaches are required to estimate dynamic errors. Moreover, the accuracy of measurements of such important flight parameters like angles of attack and glide, overload, altitude and airspeed depend greatly on the coordinates of the sensors installation, the position of the center of mass of the aircraft, the length of air pressure ducts and other technical features, as well as on the flight mode and type of maneuver. Moreover, the on-board digital recording systems, in their turn, introduce additional errors, among which the most significant are usually time shifts that occur due to inaccurate synchronization of heterogeneous information flows, as well as due to the lack of consideration for the sequence of sensors sampling within the information frame. The considered errors are dynamic and are related to the movement of the aircraft. These errors detection becomes particularly relevant, for example, in the processing of flight test data, in post-flight control of piloting near operational limits, in the flight accident investigation, etc.

2. Algorithm for accounting measurement errors

Since the above mentioned errors are clearly systematic in nature, it is advisable to apply a systematic approach for their estimation. In this paper, the relations between the flight parameters determined by the equations of spatial motion of the aircraft are used for this purpose.

From the general model of spatial motion of the aircraft [1], the following equations are selected:

$$
\begin{aligned}
&\frac{d\alpha}{dt} = \omega_z - \frac{1}{\cos\beta}\left[\left(\frac{a_x}{V} - \omega_y \sin\beta\right)\sin\alpha + \left(\frac{a_y}{V} + \omega_x \sin\beta\right)\cos\alpha\right],\\
&\frac{d\beta}{dt} = \frac{a_z}{V}\cos\beta - \left(\frac{a_x}{V}\sin\beta - \omega_y\right)\cos\alpha + \left(\frac{a_y}{V}\sin\beta + \omega_x\right)\sin\alpha,\\
&\frac{dV}{dt} = a_x \cos\alpha\cos\beta - a_y \sin\alpha\cos\beta + a_z \sin\beta,\\
&\frac{d\upsilon}{dt} = \omega_y \sin\gamma + \omega_z \cos\gamma,\\
&\frac{d\gamma}{dt} = \omega_x - tg\,\upsilon\,(\omega_y \cos\gamma - \omega_z \sin\gamma),
\end{aligned}
\tag{1}
$$

where, α, β — the angles of attack and sideslip, rad; $\omega_x, \omega_y, \omega_z$ — angular velocities relative to the associated axes, rad/s; υ, γ — pitch, roll, rad; V — actual airspeed, m/s; a_x, a_y, a_z — accelerations along aircraft associated axes defined by the formulas:

$$
\begin{aligned}
a_x &= g\,(n_x - \sin\upsilon),\\
a_y &= g\,(n_y - \cos\upsilon\cos\gamma),\\
a_z &= g\,(n_z + \cos\upsilon\sin\gamma),
\end{aligned}
\tag{2}
$$

where n_x, n_y, n_z — overloads along the associated axes, g — acceleration due to gravitation force.

The main feature of these equations is that they do not depend on the aerodynamic coefficients of the aircraft, since the accelerations included in the right parts can be calculated using the values of measured overloads in flight.

Equations (1), (2) in vector mode have the form

$$y'(t) = f\,(y(t),\, a,\, u(t)), \tag{3}$$

where $y(t), u(t)$ — vectors of output and input signals; a — vector of parameters.

In this case, the vector $y(t)$ consists of $\alpha(t), \beta(t), V(t), \upsilon(t), \gamma(t)$ signals. The vector of input signals includes the values of angular velocities $\omega_x, \omega_y, \omega_z$ and overloads n_x, n_y, n_z measured in flight.

Initial conditions $y(t_0)$ are assigned based on the results of parameter measurements $\alpha(t), \beta(t), V(t), \upsilon(t), \gamma(t)$ at the start time of the processing interval.

Direct numerical integration of the system of differential equations (1) taking into account (2) generally does not give positive results, since the constant components of the measurement errors of the input signals $\omega_x, \omega_y, \omega_z$ and n_x, n_y, n_z create linear trend-type errors at the output of the integrals, which does not allow comparing the output signals of the model and the object. To eliminate this effect, we introduce a vector of unknown parameters a consisting of constant components of measurement errors of signals $\omega_x, \omega_y, \omega_z$ and n_x, n_y, n_z:

$$a^T = \left[C_{\omega_x} C_{\omega_y} C_{\omega_z} C_{n_x} C_{n_y} C_{n_z}\right]. \tag{4}$$

To find parameter (1.4) estimates, we apply, for example, the maximum likelihood identification algorithm [2]. The observation model in vector mode has the form,

$$z(t_i) = y(t_i) + \eta(t_i), \tag{5}$$

where, $\eta(t_i)$ — the observation noise, which is a vector normal random sequence of the white noise type with zero mean and a constant covariance matrix R. The observation vector includes signals whose derivatives are on the left side (1):

$$z^T(t) = \left[\alpha(t), \beta(t), V(t), \upsilon(t), \gamma(t)\right]. \tag{6}$$

The functional to be minimized takes the form,

$$J(a) = \sum_{i=1}^{N}((z(t_i) - \hat{y}(t_i, a, u(t_i)))^T R^{-1}(z(t_i) - \hat{y}(t_i, a, u(t_i))), \tag{7}$$

where, $\hat{y}(t_i, a, u(t_i))$ — the output signal of the object model (1), (2); N — the number of samples in the processed data. The numerical optimization algorithm used to find parameter estimates is a modification of the classical Newton method and is given in [2] or [3].

To improve accuracy, the model should take into account the shift of overload sensors and sensors of angles of attack and sideslip relative to the center of mass. Corrections to overload measurements are calculated using the following formulas [4]:

$$\begin{aligned}
\Delta n_x &= \frac{1}{g}\left[\left(\omega_y^2 + \omega_z^2\right)x_0 - \omega_x\omega_y y_0 - \omega_z\omega_x z_0 + \frac{d\omega_z}{dt}y_0 - \frac{d\omega_y}{dt}z_0\right],\\
\Delta n_y &= \frac{1}{g}\left[\left(\omega_z^2 + \omega_x^2\right)y_0 - \omega_y\omega_z z_0 - \omega_x\omega_y x_0 + \frac{d\omega_z}{dt}x_0 - \frac{d\omega_x}{dt}z_0\right],\\
\Delta n_z &= \frac{1}{g}\left[\left(\omega_x^2 + \omega_y^2\right)z_0 - \omega_z\omega_x x_0 - \omega_y\omega_z y_0 + \frac{d\omega_y}{dt}x_0 - \frac{d\omega_x}{dt}y_0\right],
\end{aligned} \tag{8}$$

where x_0, y_0, z_0 — the coordinates of the sensors in the associated system with the origin at the center of mass of the aircraft. The derivatives of angular velocities are calculated numerically according to the formulas presented in [4].

When calculating the corrected overload values n_x, n_y, n_z, the corrections (8) are added to the measured values, i.e.

$$n_x = n_{x,ex} + \Delta n_x,\ n_y = n_{y,ex} + \Delta n_y,\ n_z = n_{z,ex} + \Delta n_z. \tag{9}$$

Corrections to the measured values of the actual angles of attack and sideslip due to the shift of the sensors relative to the center of mass are calculated using the formulas,

$$\Delta\alpha_{actu} = \frac{\omega_z}{V}x_\alpha - \frac{\omega_x}{V}z_\alpha + \frac{\partial\varphi}{\partial n_y}\left(n_y - 1 + \frac{\omega_x x_\alpha}{g}\right), \tag{10}$$

$$\Delta\beta_{actu} = \frac{\omega_y}{V}x_\beta - \frac{\omega_x}{V}y_\beta, \tag{11}$$

where, $x_\alpha, x_\beta, z_\alpha, y_\beta$ — the coordinates of the sensors in the associated system with the origin at the center of mass of the aircraft, m; V — actual airspeed, m/s; φ — the angle of deflection of the rod on which the angle of attack sensor is installed, degree (for on-board sensors, the last term (10) is zero).

Corrections (10), (11) are added to the actual values of the angles of attack and sideslip measured in flight, i.e.

IOP Conf. Series: Materials Science and Engineering 1027 (2021) 012011 doi:10.1088/1757-899X/1027/1/012011

$$\alpha_{actu} = \alpha_{actu\_ex} + \Delta\alpha_{actu}, \beta_{actu} = \beta_{actu\_ex} + \Delta\beta_{actu}, \qquad (12)$$

where $\alpha_{actu\_ex}, \beta_{actu\_ex}$ — the values of the actual angles of attack and sideslip calculated on board.

The considered algorithm provides calculations of constant errors of measurements of angular velocities and overloads. However, its main advantage is that it allows estimating dynamic errors of various types. Let's consider examples of processing flight test data for some modern aircraft.

3. Results of flight data processing

Figure 1 shows the results of processing a flight regime with stepwise input signal in pitch channel. The signals $\alpha_{ex}, \alpha_{\text{mod}}$ show the degree of compliance of the angle of attack measured in flight and calculated in the model (1), (2) with corrections (10). The correction value is represented separately by the signal α_{err} (digitization along the right coordinate axis). The graph shows that the value of the error caused by the non-zero coordinates of the angle of attack sensor in this mode is –0.4...0.7 degrees, which is important in precision piloting modes and near operational limits. Note that this error is not calculated on board and is not taken into account when displaying the angle of attack on the display to the pilot.

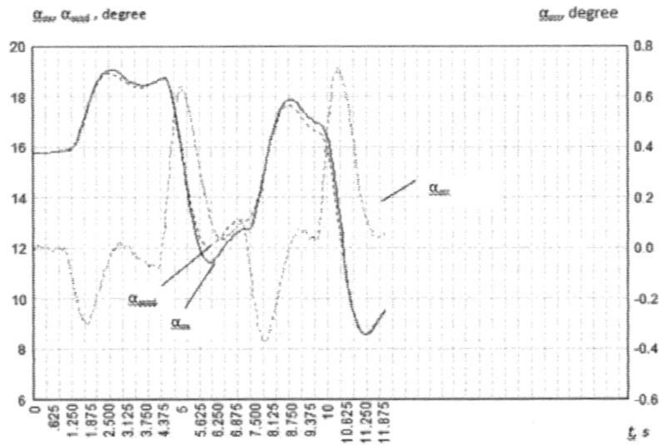

Figure 1. Estimation of measurement error of the angle of attack α_{err}, due to the sensor's shift relative to the center of mass. Comparison of angle of attack measured in the flight experiment α_{ex} and calculated in the model α_{mod}.

Figure 2 for a different type of aircraft also shows a flight regime with the stepwise deviations of the aircraft control stick in pitch channel.

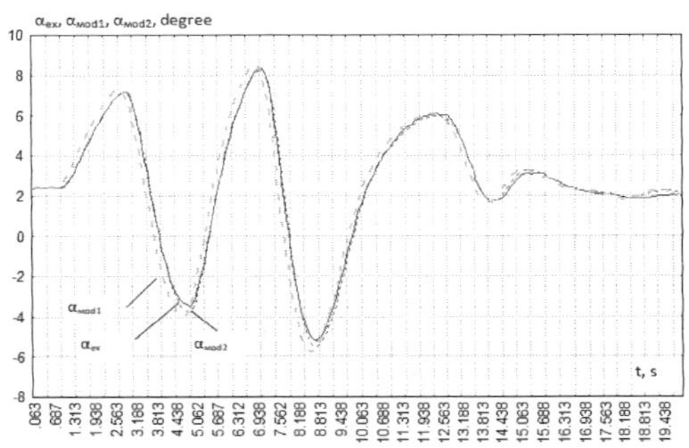

Figure 2. Detection of an error of the time delay type for measuring the angle of attack α_{ex} by comparing it with the simulated angle of attack $\alpha_{\text{mod}1}; \alpha_{\text{mod}2}$ — the simulated angle of attack after model correction.

Comparison of the flight and simulated angles of attack $\alpha_{ex}, \alpha_{\text{mod}}$ shows the presence of a dynamic error of the time shift type. Additional analysis showed that a low-frequency filter is installed on board in the angle of attack channel, which smoothens the measurement noise and creates delays of up to 0.3 s at output for registration and indication to the pilot. After taking into account the specified filter transfer function, a high degree of compliance (signals) is obtained in the model.

It is interesting to note that the detected dynamic error can lead to various errors in the analysis of flight data. For example, in flight tests and in the analysis of flight accidents, the dependence $c_y(\alpha)$ is calculated from experimental data. The values of the lift coefficient are determined by the formula, which is true under the condition $\beta \approx 0$.

$$c_y(t_i) = \frac{\left(n_y(t_i)\cos\alpha(t_i) + n_x(t_i)\sin\alpha(t_i)\right)mg - P\sin(\alpha(t_i) + \varphi_{eg})}{qS} \qquad (13)$$

Figure 3 shows the plot of the lift coefficient against the angle of attack α_{ex} measured in the flight (figure 2). The view of the graph leads to the idea of the presence of aerodynamic hysteresis. The real reason is the dynamic error discussed above. Figure 4 shows the data in figure 3 after correcting the angle of attack measurements using the filter transfer function α_{ex_corr}. As one can see, the nature of the relationship is approaching a straight line, which coincides with the aerodynamic characteristics of this aircraft. The remaining fluctuations characterize the level of non-excluded experimental errors.

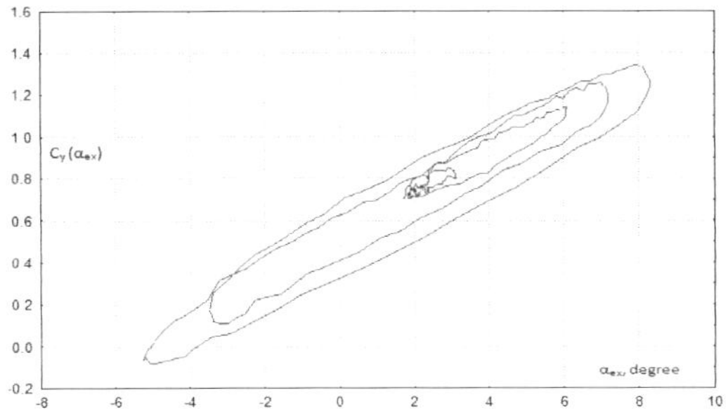

Figure 3. The effect of false aerodynamic hysteresis on the lift coefficient $c_y(\alpha)$, due to a delay in the measurements of the angle of attack α_{ex}.

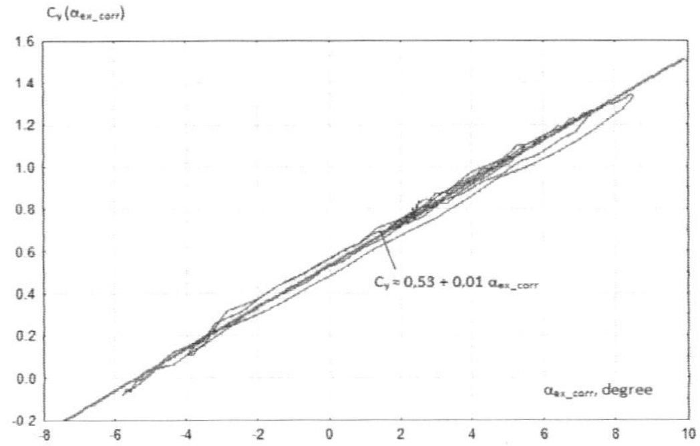

Figure 4. Refined experimental dependence $c_y(\alpha)$ after correcting the delay in angle of attack measurements.

Concluding this example we should note that actual aerodynamic hysteresis happens at much greater angles of attack, usually higher than 35 degrees. The detailed investigation of a maneuverable aircraft aerodynamic hysteresis is presented in [5].

In the case of a properly functioning on-board measurement and record system, the degree of compliance of the signals measured in flight and calculated in the model is very high, even in intensive maneuvering modes. Figure 5 shows the convergence of pitch and roll signals during the "barrel" maneuver (full 360 degrees turn in the angle of roll).

As a result of the generalized analysis of more than 20 different sections of the flight, the following values of standard deviation of mismatches were obtained: in the channels of the angles of attack and glide 0.1...0.4 degrees, the pitch angle 0.2...0.4 degrees, the roll angle 0.3...1.3 degrees, the true airspeed 0.5...0.8 m/s. Greater mismatches usually indicate that the measurement system is not working properly. So, in figure 6, deviations of the experimental values of the pitch angle from the corresponding signal of the model characterize the errors of the aircraft gyro horizon due to vigorous maneuvering in the roll channel.

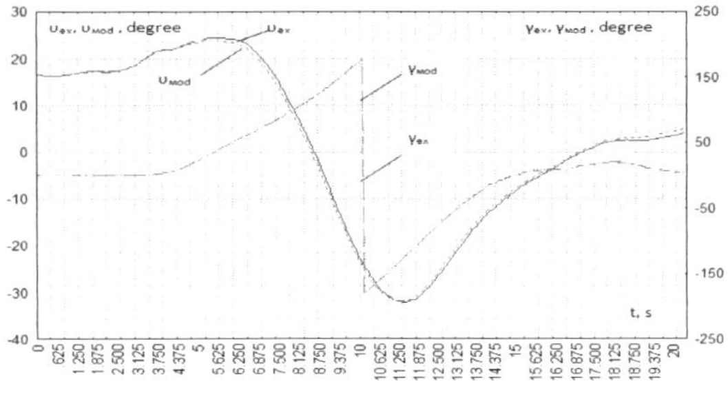

Figure 5. Comparison of pitch and roll angles measured in the experiment ($\upsilon_{ex}, \gamma_{ex}$ and estimated in the model ($\upsilon_{\text{mod}}, \gamma_{\text{mod}}$) when the on-board measurement system is working correctly. Type of maneuver – "barrel" (full roll turn).

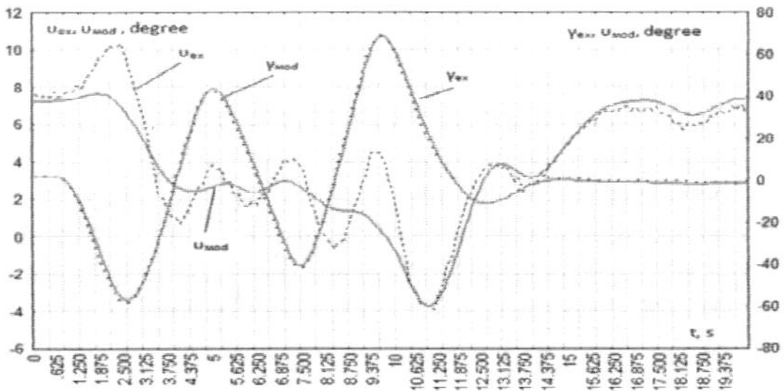

Figure 6. Detection of measurement errors of the pitch angle ϑexp by the aircraft gyro horizon due to roll movement, by comparison with the output model υ_{mod}.

Figure 7 shows experimental and simulated values of roll angles (high degree of compliance) and vertical velocity V_y (significant delay of experimental values) for the "barrel" mode. To obtain vertical velocity V_y the model (1) was supplemented with the formula,

$$V_y = V\left[\cos\alpha\cos\beta\sin\upsilon - \sin\alpha\cos\beta\cos\upsilon\cos\gamma - \sin\beta\cos\upsilon\sin\gamma\right]. \quad (14)$$

The analysis showed that the cause is the inertial properties of the barometric vertical velocity meter. For accounting them, a model of the form $K/(T_p + 1)$ was additionally introduced, the coefficients of which were included in the vector of identified parameters.

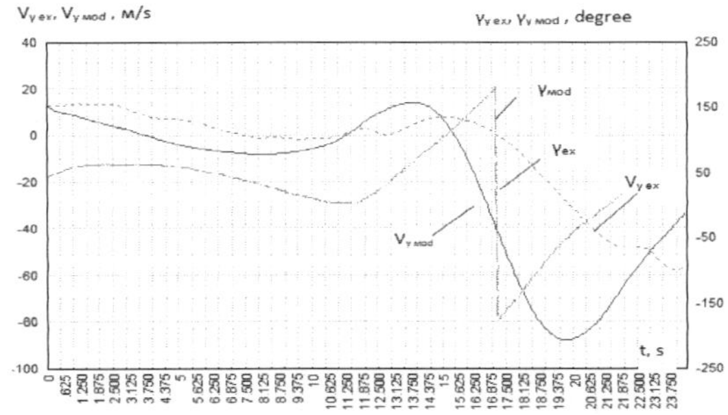

Figure 7. Delay in on-board measurements of vertical velocity V_{y_ex} due to the inertial properties of the barometric measuring device. Type of maneuver — "barrel".

The result is shown in figure 8, which confirms that the introduced simple model takes into account the main components of the dynamic measurement error in V_y channel.

This approach also makes it possible to detect delays in the true airspeed measurement channel. Figure 9 shows a high correspondence between the simulated and experimental speed values for the stepwise input signal in yaw channel, which was obtained after correcting the experimental data by the time shifting to advance by an amount of 0.6 s. The angle of sideslip channel in this case required no correction.

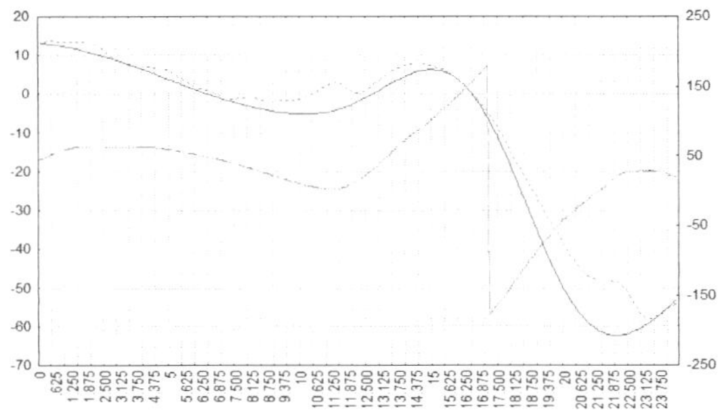

Figure 8. Coincidence of the vertical velocity values measured in the flight experiment V_{y_ex} and the output signal of the model V_{y_mod} after the model was supplemented with a first-order inertial filter $K/(T_p+1)$.

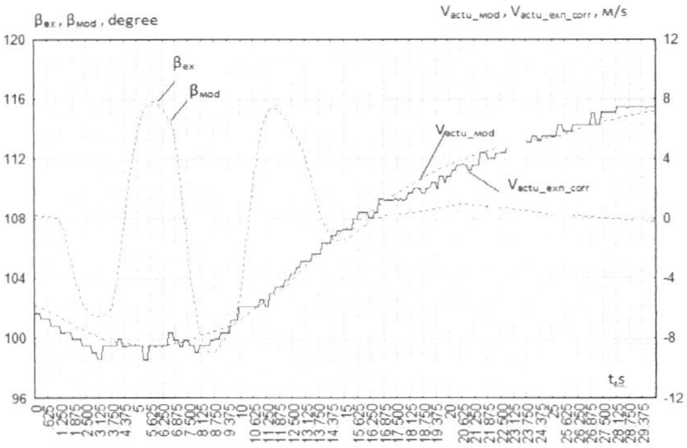

Figure 9. Coincidence of the values of the sideslip angles measured in the experiment and calculated in the model with actual airspeed after correcting the speed measurements.

4. Conclusion

Thus, the considered approach makes it possible to detect various measurement errors of some basic parameters that are essential for flight safety issues and in during post-flight analyses. It should be noted that the idea of using a motion model of type (1) to check the correctness of onboard measurements has been repeatedly expressed earlier in some publications, for example, [2, 4, 6]. The originality of the material presented above consists in a specific algorithmic implementation, selection of a set of identifiable parameters, and determination of the types of estimated errors. In addition, the results are in good agreement with those obtained earlier using identification methods [5, 7].

Acknowledgments

The paper is supported by Russian Foundation for Basic Research, project 20-08-00449.

References

[1] *Aerodynamics, stability and controllability of supersonic aircraft* 1998 ed G S Bjushgens (Moscow: Nauka) p 816
[2] Klein V and Morelli E A 2006 *Aircraft system identification: Theory and practice* (Reston, VA: AIAA, Inc.)
[3] Bulgakov V V, Korsun O N, Kulabukhov V S, Stulovskii A V and Timofeev D S 2016 *Journal of Computer and Systems Sciences International* **55** (1) 150–61
[4] Vasilchenko K K et al *Flight tests of aircraft* 1996 (Moscow: Mashinostroenie) p 720
[5] Korsun O N, Stulovskii A V, Ovcharenko V N and Kanyshev A V 2018 *Journal of computer and systems sciences international* **57** (3) 374–89
[6] Jategaonkar R V 2006 *Flight vehicle system identification: A time domain methodology* (Reston, VA: AIAA, Inc.)
[7] Nikolaev S V and Korsun O N 2019 Nonlinearities evaluation in the aircraft parameter identification *IOP Conference Series: Materials Science and Engineering* **476** 012020

Speech spectrum analyses for estimating operator functionality

O N Korsun[1,2], V N Yurko[1], S A Beyseev[3], A S Naukenova[3] and A K Tulekbaeva[3]

[1] State Research Institute of Aviation Systems,7 Viktorenko, Moscow 125319, Russia
[2] Moscow Aviation Institute (National Research University), 4 Volokolamskoe Shosse, Moscow 125993, Russia
[3] M. Auezov South Kazakhstan University, Shymkent 160012, Kazakhstan

marmotto@rambler.ru

Abstract. The paper deals with the problem of estimating human operator functional characteristics using frequency analysis of operator's speech. The analysis is carried out using operator's speech transfer function based on the classical definition of transfer function in the automatic control theory. The algorithms for calculating the operator's speech transfer function are presented, the experimental results are discussed.

1. Introduction

In recent years, there has been an intensive development of man-machine interface tools in technology, such as modern audio technologies [1–3], neural networks [4] etc. In [5], the correlation between human operator physiological state and variations of his speech are investigated.

This work is devoted to the analysis of speech signal in the frequency domain for the purpose of estimating operator functionality. The analysis is based on the speech transfer function, which is introduced by analogy with the transfer function of the classical theory of automatic control [6].

2. Algorithm for calculating a frequency transfer functions

Let us consider an audio record of a word and divide it into intervals of 20–40 ms. Every interval contains a sequence $x(N)$, where is N — the number of samples. Then the following operations are performed [3, 7]:

- signal pre-filtering

$$x(n) = x(n) - \alpha \cdot x(n-1),\ \alpha = 0.95;$$

- signal frame extraction

$$s_m(n) = x(m \cdot \Delta n + n),\ \ 0 \le n \le N_{FFT} - 1,$$

where $\Delta n = N_{FFT} - \varepsilon \cdot N_{FFT}$, N_{FFT} — length of fast Fourier transform in samples (frame length); ε — the ratio of the length of the overlap section to the length of the frame, $0 \le \varepsilon \le 0.5$; m — frame number;

- frame weighting by Hanning window

$$w_{H2} = 0{,}5\left(1-\cos\left(\frac{2\pi n}{N_{FFT}-1}\right)\right) \text{ — Hanning window;}$$

$$s_m(n) = w_{H2}(n) \times s_m(n);$$

- calculating the fast Fourier transform and the square of its modulus, or estimating the spectral density

$$X_m(k) = \text{FFT}\{s_m(n)\} = \sum_{n=0}^{N_{FFT}-1} s_m(n) \cdot \mathrm{e}^{\frac{j\cdot 2\pi k n}{N_{FFT}}},$$

$$S_m(k) = |X\_m(k)|^2 = Re(X_m(k))^2 + Im(X_m(k))^2,$$

- calculation of the average spectral density in the spectral band

$$S(f_i) = \frac{1}{\Delta k}\sum_{k=k_{i,\min}}^{k_{i,\max}} S_m(k),\ i = 1, \ldots, N_{frb},$$

where $\Delta k = k_{i,\max} - k_{i,\min} + 1$ — spectral band width; $k_{i,\max}, k_{i,\min}$ — indices of the first and last frequency in the spectral band,

$$k_{i,\min} = \left\lfloor \frac{N_{FFT}}{f_s}\left(f_{\min} + \frac{f_{\max} - f_{\min}}{N_{frb}}(i-1)\right)\right\rfloor + 1,\quad i = 2 \ldots N_{frb},\quad k_{1,\min} = \left\lceil \frac{N_{FFT}}{f_s} f_{\min}\right\rceil,$$

$$k_{i,\max} = \left\lfloor \frac{N_{FFT}}{f_s}\left(f_{\min} + \frac{f_{\max} - f_{\min}}{N_{frb}} i\right)\right\rfloor,\quad i = 1 \ldots N_{frb},\ f_i = f_{\min} + \frac{f_{\max} - f_{\min}}{N_{frb}}\left(i + \frac{1}{2}\right),\quad i = 1 \ldots N_{frb},$$

f_i — frequency corresponding to the middle of the *i*-th band; f_S — sampling frequency;

Thus, on each time interval of 20 ... 40 ms duration, the speech signal is considered as a sequence of spectral density values averaged over the frequency band. This approach is based on the classical definition of the spectral density of the signal [6].

The algorithm described above is applied to each word realization, as a result, we obtain a matrix of dimension $N_{frb} \times N_t$, in which the number of rows N_{frb} is equal to the number of spectral bands, and the number of columns N_t is the number of time quantization intervals. Normally the number of spectral bands $N_{frb} = 35$, the number of time intervals $N_t = 17$.

Consider calculating the transfer function between two speakers to reveal the differences in frequency properties.

Let the first speaker say N realizations of a word.

The average matrix is calculated

$$X_{D1} = \frac{1}{N}\sum_{k=1}^{N}\{x_{ij}\}_{D1};\ i = 1, \ldots, N_t;\ j = 1, \ldots, N_{frb},$$

where N_t — is the number of time intervals into which the considered word record is divided; N_{frb} — the number of spectral bands.

The matrix for the second speaker X_{D2} is calculated similarly. If it is required to find changes in speech properties between different states of one speaker, then the X_{D2} matrix is calculated for the second state of this speaker.

When calculating the transfer function, we additionally carry out averaging for each frequency band over N_t time intervals. Finally, we get the formula for the modulus of the transfer function between the first and second speakers in decibels (dB):

$$W_{12}(f_j) = 10\log_{10} \frac{\frac{1}{N_T}\sum_{i=1}^{N_t}\{x_{ij}\}_{D1}}{\frac{1}{N_T}\sum_{i=1}^{N_t}\{x_{ij}\}_{D2}}.$$

3. Dependence of the estimates of the speech transfer function on the structure and volume of speech material

To test the performance of the above algorithm, the following experiment was performed. Under normal conditions, the speaker pronounced the Russian words "пилотаж" (aerobatics), "масштаб" (scale), "навигация" (navigation) in 50 realizations of each. Note that the analysis of the above algorithm suggests that it does not depend on the language used. Indeed, in [7] for a similar algorithm using examples of Russian, Kazakh and English languages, it was shown that changing the language does not affect the accuracy of automatic word recognition.

The experiment was repeated after 4 hours, also under normal conditions. Based on the data obtained, the transfer functions were calculated separately for each word for all its implementations (Figure 1) As you can see, the values of the function basically correspond to a unity gain (level 0 dB), which indicates the constancy of the speaker properties. In this case, deviations of this do not exceed ± 1.5 dB. Figure 1 also characterizes the degree of dependence of estimates on the structure of speech material: deviations of estimates calculated for individual words relative to the average do not exceed ± 1 dB.

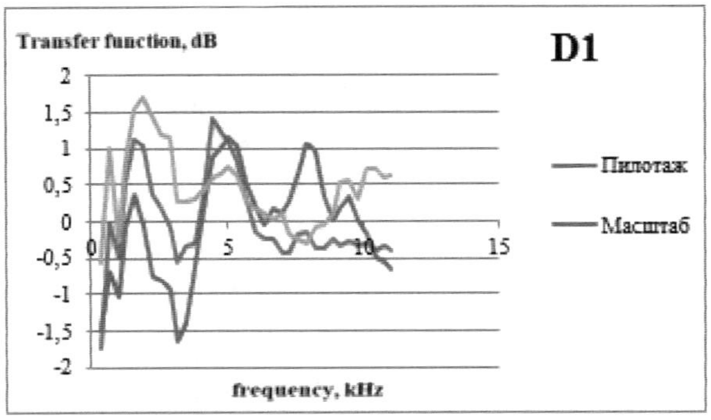

Figure 1. Graph of the speech transfer function for the speaker D1 according to the speaker D1 in normal conditions with an interval of 4 hours between recordings.

To establish the dependence of the estimates averaged over three words on the number of realizations, we compared the values of the function calculated over 50 realizations and taken as a standard with estimates for n = 3, 5, 10, 15, 20, 25 realizations. The standard deviations (s.d.) σ_i of each estimate from the standard were considered as a measure of the error (Table 1). The table shows that already with 3 realizations the s.d. of errors do not exceed 0.5 dB, which is sufficient to detect significant changes.

Table 1. Standard deviations of errors in the estimates of the speech transfer function depending on the volume of speech material (number of realizations).

N	3	5	10	15	20	25
σ_i, dB	0.492	0.395	0.212	0.179	0.092	0.049

Let us further consider several examples characterizing the capabilities of the proposed speaker's transfer function for detecting changes in speech in various conditions. In these examples, to eliminate the influence of sample size, estimates are calculated for 50 realizations of each word.

4. Influence of noise-protective headphones on speech

This section deals with the transfer functions for speakers in normal conditions, and for the same speakers in noise protection headphones.

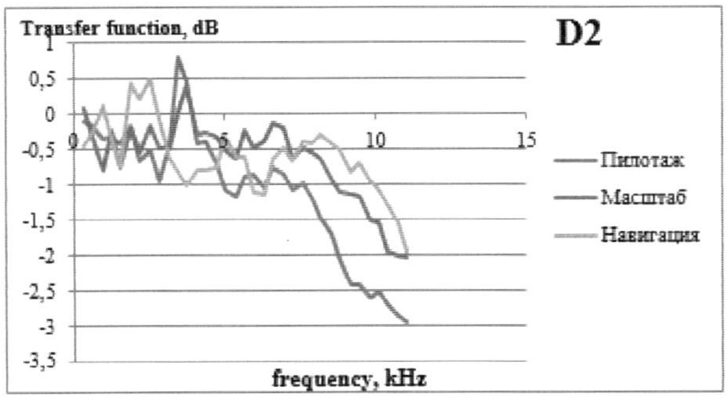

Figure 2. Graph of the transfer function for the speaker D2 in noise-protective headphones.

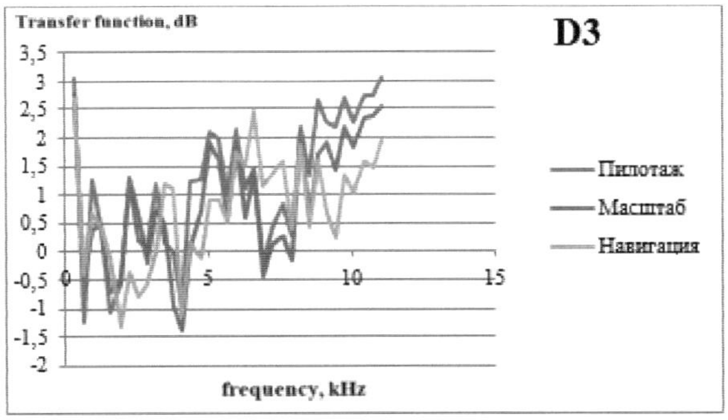

Figure 2. Graph of the transfer function for the speaker D2 in noise-protective headphones.

The resulting graphs show the effect of noise protection headphones on the speaker. It should be noted that the reactions of different speakers differ from each other, however, typical is the change in amplitude up to ± 3 dB in the frequency range 6 ... 11 kHz.

5. Impact of noise on the state of a speaker over time

Consider the issue of changing the speaker's state during the experiment. To do this, a set of 150 words was divided into 4 parts, after which the transfer function for each part was calculated. The total time of the experiment was 2 minutes, the duration of one part was 0.5 minutes. During the experiment, the speaker was exposed to a low-frequency noise of 90 dB in the frequency band up to 2 kHz.

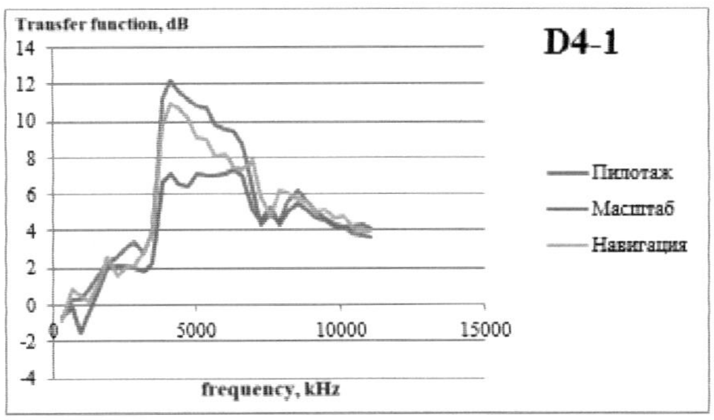

Figure 4. Graph of the transfer function for the D4 speaker in low frequency noise conditions, part 1.

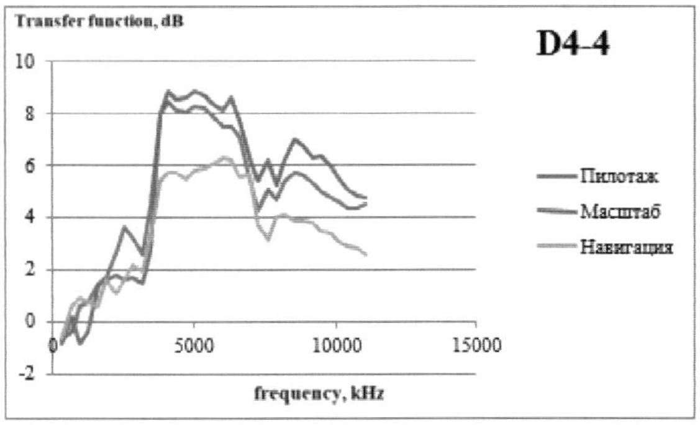

Figure 5. Graph of the transfer function for the D4 speaker in low frequency noise conditions, part 4.

In general, the graphs for each part are similar, but the amplitude of the graphs of the 4th part is lower. This can be explained by the fact that over time the speaker gets used to the noise and the volume of his voice decreases, that is, the Lombard effect weakens (Lombard effect — the instinctive rise of the speaker voice volume in noisy environment [1]).

6. The influence of hearing defects on the frequency properties of speech

Let us consider the application of the proposed function to identify the speech features of a group of helicopter pilots who have an occupational hearing disorder — hearing loss. A speaker who had no diagnosed diseases of hearing and speech was taken as a reference speaker. The results for the three speakers in this group are shown in Figure 6–8.

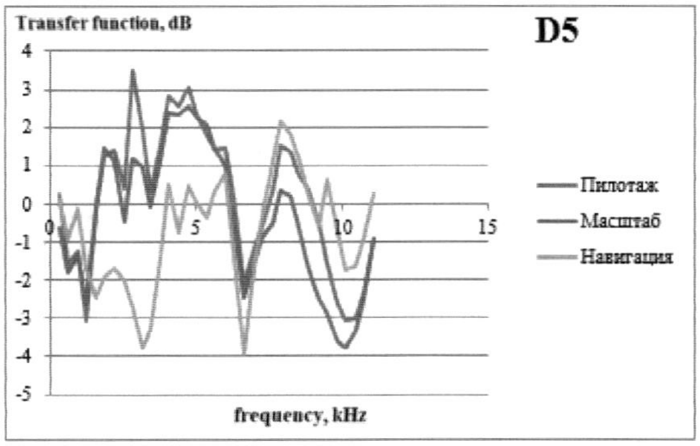

Figure 6. Graph of the transfer function for the speaker D5.

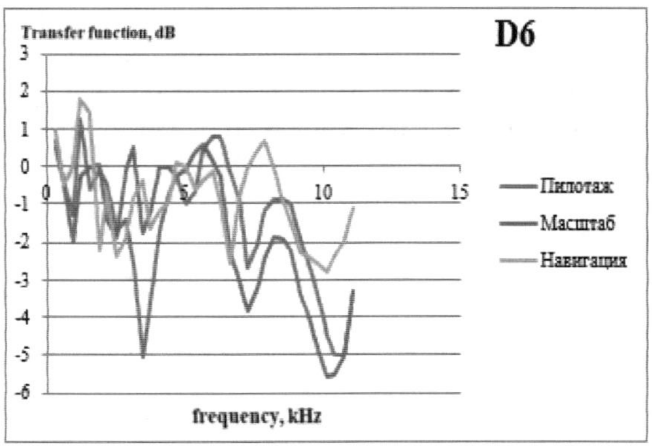

Figure 7. Graph of the transfer function for the speaker D6.

Comparison with the transfer function of a speaker without hearing diseases (Figure 1, the spread is within ± 1.5 dB), shows that the graphs of the transfer functions for this group of speakers differ in a greater variation of values: for speaker D5 from –4 to 3 dB, for speaker D6 from –5 to 2 dB, for speaker D7 from –6 to 4 dB. Thus, the method considered in the article provides a good diagnostic feature for identifying speakers with hearing diseases based on their speech samples.

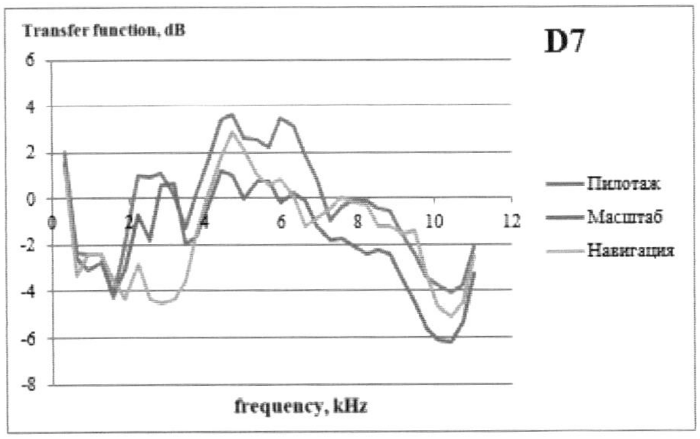

Figure 8. Graph of the transfer function for the speaker D7

7. Conclusion

The paper reports the following main results:

- an algorithm for calculating the speaker's speech transfer function, which enables to analyze changes in the speech frequency spectrum due to the external conditions or the speaker's state;
- examples are presented that characterize the capabilities of the proposed speech transfer function to detect the state of the speaker external conditions or.

Acknowledgements
This work was supported by Russian Foundation for Basic Research (RFBR), project 18-08-01142-a.

References

[1] Rabiner L and Juang B H 1993 *Fundamentals of Speech Recognition* (Englewood Cliffs, New Jersey: Prentice Hall International, Inc.) p 507
[2] Bondaros Yu G, Ivanov A I and Tischenko A A 2010 Operator fatigue degree definition according his voice Lyapunov exponent *Herald of computer and information technologies* **6** 22–30
[3] Korsun O N and Poliev A V 2016 Automated definition of phonetically homogeneous sections of words in a natural language based on multiparameter optimization *Journal of Computer and Systems Sciences International* **55** (4) 609–18
[4] Evdokimenkov V N et al. 2015 The use of the neural network model of pilot control actions for their individually-adapted support *Journal of Computer and Systems Sciences International* **54** (4) 609–20
[5] Ivanov A I. et al. 2014 Study of speech characteristics in hearing-impaired pilot speakers for creation of voice-activated system operating airborne equipment *Occupational Health and Industrial Ecology* **11** 40–5
[6] *Methods of classical and modern theory of automatic control* 2004 ed K A Pupkov and N D Egupov (Moscow: Izd. MGTU im. Baumana) p 656
[7] Korsun O N, Gabdrakhmanov A Sh, Mihajlov E I, Nahaev M Z and Tulekbaeva A K 2015 Algorithm for an automatic recognition of the speech commands, invariant to languages *Mekhatronika, Avtomatizatsiya, Upravlenie* **16** (9) 599–604

IOP Conf. Series: Materials Science and Engineering 1027 (2021) 012013 doi:10.1088/1757-899X/1027/1/012013

Digital recruitment of innovative project teams' leaders based on genetic algorithms

E E Kovshov[1], V S Kuvshinnikov[1] and L E Osipenko[2]

[1] JSC"NIKIMT – Atomstroy", Moscow 127410, Russia
[2] Moscow City University, Moscow 129226, Russia

KovshovEE@atomrus.ru

Abstract. Over the past decade, innovation has become an important part of the development for large business companies. The search for promising topics, as well as the recruitment of potential applicants teams who can implement an innovative project with the maximum benefit for business at the lowest cost, led to the search for approaches to solving these problems using IT. Since the problems of such research are covered separately in different industry areas, we implemented an approach to completing project teams based on genetic algorithms. A wide review of the literature allowed us to identify some important characteristics of innovative projects, as well as to summarize the portfolio of competencies required for potential designers. These attributive characteristics of the project and the applicants' competence should be taken into account. The developed genetic algorithm allows searching for multiple parameters simultaneously and generating new solutions through the use of different operators. The results open up a new range of opportunities for business companies and government organizations interested in successful implementation of innovations.

1. Introduction

Currently, the economic performance of business companies and their ability to retain competitive industry advantages depend on their innovation potential and innovation activity [1-3]. Innovation management is a rather complex and multi-criteria task, since it includes an analysis of a significant number of such heterogeneous elements as human resources, financial and technological components of an innovation project (figure 1).

One of the central components of innovation management is the workforce. Postulating the thesis "Cadres are all-important", owners of business corporations adequately assess not only the importance of forming a cohesive, efficient project team, but also the search for alternative options of "efficiency of designers — quality of innovative product" balance. The latter suggests that other experts, subject your material to such powerful checks that they can bring the race of evidence to the point where all your resources may not be enough to win [4].

This kind of "battle" increases the cost of proof race, multiplies the number of tests, changes the shape of new objects, thereby actualizing the need to find solutions that can track and analyze set of indicators and parameters described by big data and processed by complex algorithms.

A wide range of identified problems leads to the formulation of three important tasks: justification of the main attribute characteristics of an innovative project and a team of designers, as well as ways to manage them based on IT solutions.

We will reveal each of these tasks sequentially.

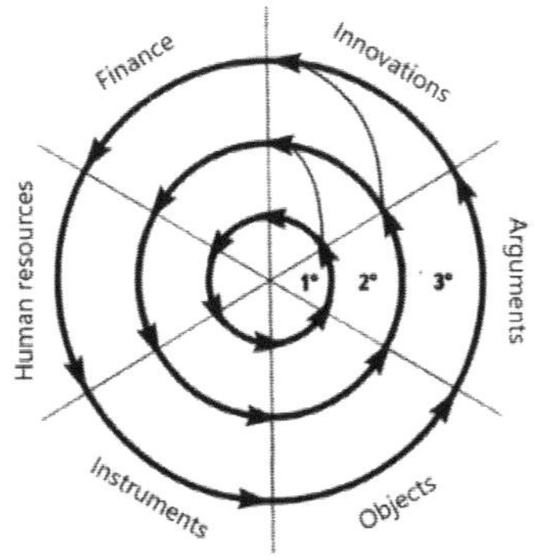

Figure 1. Structural components of the innovation management process [4].

2. Attribute characteristics choice

Speaking about the search of ideas for innovative projects by business companies, it should be noted that a characteristic feature of modern post-industrial society is the "transformation of science into technoscience". In the symbiosis "fundamental science ↔ technology → wealth of society", science is a generator of new ideas and principles for high-tech developments. Technology, in turn, creates new products and solutions that expand human capabilities [5].

The "applied science" trend has led to an increase in the part of technological developments in total scientific research volume, and as a result — the possibility of their typology based on the subject area nature into research, design and complex.

The commercialization of science and the gradual replacement of fundamentals with the criterion of efficiency and practical utility — determine the range of an innovative project targeted results, which varies from supporting, involving the improvement of existing knowledge and solutions, to universal and breakthrough.

Traditional typology based on the execution duration implies the presence of short - term, medium-term and long-term projects.

These main attribute characteristics of the innovation project led to the following discursive field for discussion of the project team. Formed like theater companies that are created temporarily for the production of a play, such project teams gather "performers" of different roles. The diversity principle assumes that the scientific productivity of each member of the project team will increase with the increase of its functions and specialties, generating nonlinear interaction between them and, as a result, not additive, but a more dramatic increase in the dependence of the collective result on the efforts of individual specialists [6, p. 226].

In this regard, we consider generalizing the main attributes of an innovative project team as a system object is important. The controlling mechanism of such team is the motivation of designers to get the maximum benefit for business while minimizing costs.

K. O. Sokolov identifies external and internal factors that ensure the effective work of the innovative project team. Among other things, he highlights managerial and organizational components, such as the employment of an applicant in other projects as a leader, expert, performer, as well as the assessment of the professional, personal and psychophysiological potential of members of the innovation team [7].

We believe that the functioning of an innovative project team will be successful if it has one or more specialists with an academic degree who are responsible for the formulation of problems, the construction of research design, as well as having sufficient authority and experience to attract the most effective designers [8].

Practice shows that in some cases the success of an innovative project is determined by the participation of a frontman; a designer with special competencies, such as technical or industrial expertise, etc. [9].

Group projects share the need for finding efficient ways to improve the effectiveness of communication between designers, for example, when discussing complex issues related to project design, clarifying and adding incomplete requirements.

For breakthrough innovation projects, it is important to justify the involvement of single inventors sitting in the "ivory tower", such as Edison or Wright.

Since the third-party IT specialists involvement is usually very expensive, and, like any designer, they have both pros and cons, and their performance in a particular innovative project depends on a variety of components, it is important to offer some objective abstract metric of "coolness" for them, too. We have accepted the main programming statuses as such attribute characteristics: Junior, Middle, and Senior.

The main professional requirement for a junior is the ability to independently and quickly perform the technical tasks assigned to him or her. In addition to energy and determination, it is important for juniors to have a desire to learn and develop, as well as the ability to calmly treat criticism.

The main requirement for a middle developer is an understanding of business requirements and the ability to translate them into specific technical solutions.

The highest-status IT specialist is senior. They should not only have extensive experience, a deep understanding of the libraries and frameworks structure, but also be able to assess the technical risks of the project. Senior has developed communication skills that allow him to recommend constructive ways to refactor code, as well as successfully convince the customer and the team. An example of an attributes set of an innovative project and its potential performers is shown in table 1.

Table 1. Example of an an innovative project and its potential performers attributes set with values.

Attribute	Possible attribute values	Numerical value	Relative attribute weight
	Innovative project		
Classification	Research	1	1
	Project	2	
	Complex	3	
Term of execution	Short-term	1	2
	Medium-term	2	
	Long-term	3	
Target result	Supporting	1	3
	Universal	2	
	Groundbreaking	3	
	Potential candidate		
Education	Secondary	1	1
	Higher	2	
	Academic degree	3	
Current employment in projects	3 and more	1	2
	2	2	
	1	3	
Special innovative competences	No	1	2
	Yes (common)	2	
	Yes (project specific)	3	
Experience in innovative development	Separated	1	3
	Single (1–2 projects)	2	
	Significant (3 and more projects)	3	
«Teamwork»	No (individual contributor)	1	3
	Equally	2	
	Always	3	
Programmer status	Junior	1	3
	Middle	2	
	Senior	3	

3. Genetic algorithm building methodology

As noted by a researcher at the Oxford Internet Institute and The Institute of Turing's Sandra Wachter, the so-called human resources analytics (people analytics) plays an increasingly important role in companies 'decisions on hiring, firing, evaluating performance and improving employees. [10]. However, it should be noted that the process of finding an equation that describes the optimal behavior of any natural system is not a simple task. According to Hoda Lipson, previously, it could have taken a scientist a lifetime to create one such optimization model [11].

Currently, new paradigms are used to solve optimization problems, which search for a balance between efficiency and quality of solutions due to the "survival of the strongest alternative solutions" in uncertain and fuzzy conditions [12–15] et al. One of these paradigms is genetic programming. M. Schmidt expressed its possibilities very vividly, noting that physicists (Newton or Kepler) could run a genetic algorithm on a computer and, after just a few hours of calculations, get laws explaining the fall of an apple or the movement of planets [16].

Genetic programming repeats the principles of biological evolution. The genetic algorithm (hereinafter referred to as GA) starts composing equations by randomly matching different building blocks consisting of mathematical expressions, and then checks how well the resulting equations describe the data. Equations that do not pass verification are rejected, and those that demonstrate a certain potential are recomposed in such a way that they can eventually turn out to be an accurate mathematical model [17].

GA is much more effective than traditional "regression" in some cases. For example, in a regression search, the form of the equation (dependency) is set in advance, and the parameters of the equation are optimized to match the data. However, in some cases it is necessary to search for the equations themselves, including arithmetic operators, trigonometric and logarithmic functions, constants, etc.

The effectiveness of GA for improving the quality of digital recruitment management of innovative project teams is determined by such features as: working with a coded set of parameters; simultaneous search among several alternatives on a given set of solutions; using the target function (hereinafter referred to as TF) to evaluate the quality of solutions; and a probabilistic approach to optimization problems analyzing.

Traditionally, the algorithm for evaluating of innovative project team members' potential is a step-by-step procedure described below. It starts with the selection of characteristics that define the innovative potential of teams; selecting the appropriate methods of assessment; development of evaluation scenario; evaluation processing; interpretation of results; wording of conclusions; developing recommendations on the team acquisition and development (figure 2).

For the GA to work, select a natural parameters set of the optimization problem and encode them into a sequence of finite length in a certain alphabet. The algorithm cycle is repeated until the specified number of algorithm iterations is completed or a required quality solution is obtained on some generation, or if a local optimum is found and premature convergence occurs, when the algorithm cannot find a way out of the local extremum influence area.

In the team selection management optimizing process, the parameter space under study is quite large, and the task does not require finding of the exact global optimum. It is enough to find an acceptable solution from a set of the most suitable ones in a given short time. In this regard, it is most appropriate to use GA at stage 3 (figure 2). The first step is to determine the initial data of the specific innovation project designer selecting process (table 2).

Each object (any vector element from table 1) can be represented as a set of attributes that characterize this object numerically. Attributes are defined on a limited set of positive values. Selection vector: $SV = \{F_1, F_2, ..., F_I\}$, where F_I — frontman number, selected for i-th innovative project, innovative projects $I = \{1, ..., N_I\}$, frontman candidates $F = \{1, ..., N_F\}$.

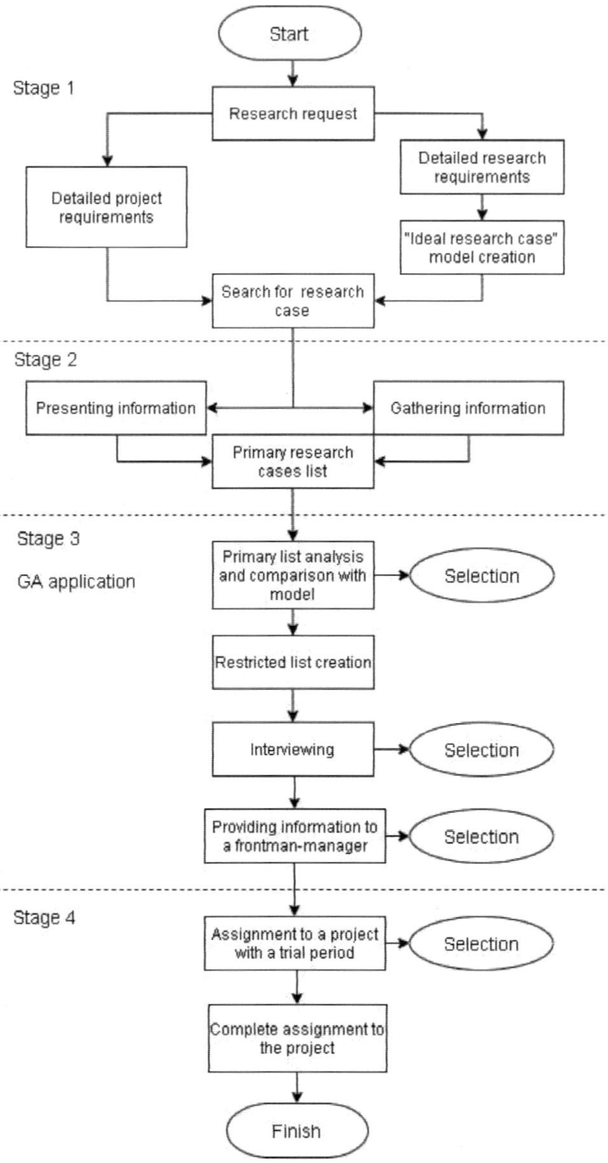

Figure 2. Selection sequence of potential designers for an innovative project implementation.

Table 2. Source data for the genetic algorithm.

1.	Potential frontman	$F = \{1, ..., N_F\}$, N_F — number of potential front-runners
2.	Innovative project	$I = \{1, ..., N_I\}$, N_I — number of innovative projects

Thus, the problem of selecting a potential candidate for a particular innovative project performance reduces to the equivalent problem of finding such an SV of candidates which will best contribute to achieving the goal considering given limitations and criteria.

The number of innovative projects should not exceed the number of potential designers. One frontman should be assigned to a specific innovation project. The degree to which the frontman-candidate's attributes (FA) match the project's attributes (PA) is estimated by converting them to universal attributes (UA) using a matching table. The table is formed depending on the specific case and allows many-to-many relationships. The attribute value is reduced to a scale of universal attributes that contains integers from 0 to 9. Depending on the difference between the corresponding attributes of the candidate and the project, a specific value is taken from a separate table of relative efficiency constants (REC). The value measures the potential compatibility of the candidate and the project.

For a more accurate assessment of the assignment effectiveness, the amount of work (laboriousness) for each of the project attributes (PL) is taken into account separately. The approximate target satisfactory amount of costs for the completion of the i-th project is defined by the formula (1):

$$\mathbf{SC}_i = \sum_{k=1}^{\mathrm{N_A}} \left(\frac{\mathbf{PL}_{i,k}}{REC(\mathbf{RUAF}_{i,j,k} - \mathbf{UAP}_{i,k})} \right) + CA_i, \tag{1}$$

where i — project number from $\{1, ..., N_I\}$; j — number of a frontman candidate; N_A — amount of attributes, k — number of a universal attribute; $\mathbf{PL}_{i,k}$ — complexity of i-th object's k-th attribute; $\mathbf{RUAF}_{i,j,k}$ — level of the j-th candidate's k-th universal attribute, required for the i-th project's implementation; $\mathbf{UAP}_{i,k}$ — the i-th project's k-th universal attribute; *REC()* — function of relative effectiveness, which values for each attribute are pre-recorded; CA_i — custom individual coefficient for the i-th project, allowing specific implementation requirements consideration. The *CA* coefficient allows you to adjust the project's requirements for the frontman's abilities, due to increased urgency, execution quality requirements, or other requirements that increase the amount of resources required.

The expected project cost of the i-th project is defined by the formula (2):

$$\mathbf{PC}_i = \sum_{k=1}^{\mathrm{N_A}} \left(\frac{\mathbf{PL}_{i,k}}{REC(\mathbf{UAF}_{j,k} - \mathbf{UAP}_{i,k})} \right) + CW_j, \tag{2}$$

where i – project number from $\{1, ..., N_I\}$; j — number of a frontman candidate; N_A — amount of attributes, k — number of a universal attribute; $\mathbf{PL}_{i,k}$ — complexity of i-th object's k-th attribute; $\mathbf{UAF}_{j,k}$ — k-th universal attribute number of j-th frontman candidate; $\mathbf{UAP}_{i,k}$ — k-th universal attribute number of i-th innovative project; *REC()*— function of relative effectiveness; CW_j — estimated current workload of j-th candidate. The *CW* factor represents the additional resource costs associated with assigning an already employed candidate to the project.

Each specie will be represented by a chromosome when initial population of GA is created. A chromosome is a selection vector ***SV***, containing N_I randomly ordered numbers of applicants according to the number of projects. Since a candidate can only be assigned to one project at a time, each gene (candidate's number) must occur no more than once in any relevant chromosome. This type of "enumerable chromosomes with unique genes" is often used in combinatorial problems. The standard crossing operator is incorrect for this type of chromosome, so a more complex two-point crossing scheme is used.

The effect of frontman candidates assigning is estimated for each selection vector separately using this equation (3):

$$\mathbf{AE}_g = \sum_{i=1}^{\mathrm{N_I}} LOSS\left(\mathbf{SC}_i - \mathbf{PC}_i\right), \tag{3}$$

where g — number of selection in population; N_I — amount of projects; k — universal attribute number; *LOSS()*— loss function, which defines the amount of resource losses caused either by a lack or an excess of resources assigned to the project. Usually, the lack of resources is more dangerous than excess, because it marks the lack of specialist's experience, leads to additional delays, decrease in quality, and may cause failure to meet deadlines. However, assigning an overly valuable specialist to a simple project leads to indirect losses associated with overpayments for professionalism or an unreasonable risk of a temporary experienced personnel shortage if more complex projects are to appear.

To determine and implement the GA we define the following concepts.

Target function — numerically represents the result of frontmen selection. Corresponds to the effect of assigning AE.

Population — a set of ***SV*** vectors.

Population size — total number of ***SV*** vectors. The population size is set before the GA starts. It remains constant during the entire calculation process.

Specie — single ***PV*** vector.

Gene — part of ***PV*** vector.

GA stop condition: receiving a solution of the required quality; GA getting into the deep local optimum of the target function; specified search time expiration.

The next step involves defining characteristics and values of attributes (table 2) that determine candidate's and project's unified attributes' weights.

The GA operation sequence used for finding an optimal solution to the frontmen recruiting problem is shown in figure 3.

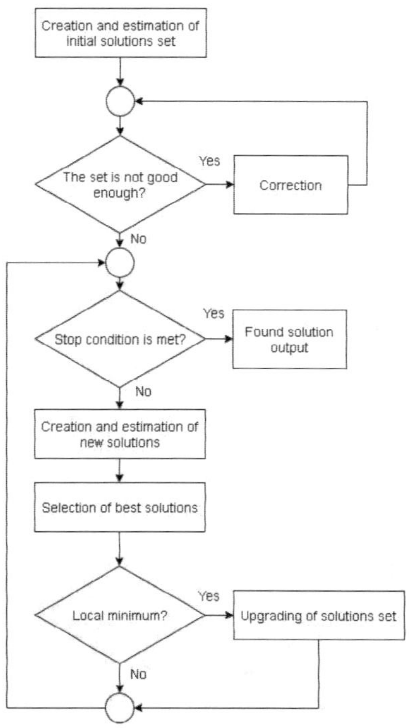

Figure 3. The genetic algorithm flowchart.

The given GA model is software implemented in the Jupiter Lab v2 programming environment using freely distributed libraries, modules and software components, mainly open source.

4. Conclusions

Thus, the interaction of pedagogy and IT can enrich and expand both sciences in the field of innovative project teams recruitment. Their pedagogical analysis as system objects, supplemented with the capabilities of genetic programming, together allows us to achieve the emergent properties of innovative project teams, maximizing business benefits while minimizing costs. Clearly aware of the complex difficulties and limitations associated with picking innovative project teams using genetic algorithms, it is necessary to emphasize the possibility of extrapolation to any level of projects implementation, as well as to the rational and emotional spheres of applicants in digital recruitment.

References

[1] Davidson N, Mariev O and Pushkarev A 2018 A Regional factors of innovation activity of Russian enterprises *Foresight* **12** (3) 62–72
[2] Kashparov D and Kochin M 2017 From startup training as a training tool for developing innovations to the innovative economy of the region and country *Bulletin of Immanuel Kant Baltic Federal University* **4** 52–9
[3] Kilov D 2018 Improving the efficiency of innovative activity of companies *Foresight* **12** (3) 30–3
[4] Latour Bruno 2013 *Science in action: How to Follow Scientists and Engineers through Society* (St. Petersburg: publishing house of the European University in St. Petersburg) p 414
[5] Mamchur E 2010 Fundamental science and technology: search for mechanisms of interaction *Modern Technologies: Philosophical and Methodological Problems* https://iphras.ru/uplfile/natsc/articals/mamchur/modern-tehnology.pdf
[6] Yablonsky A 2001 *Models and methods of research of science* (Moscow: Editorial URSS) p 400
[7] Sokolov K 2011 Evaluating the potential of an innovative team *Digest-Finance* **8** https://cyberleninka.ru/article/n/otsenka-potentsiala-innovatsionnoy-komandy
[8] Peltz D and Andrews M 1973 *Scientists in organizations. About optimal conditions for research and development* (Moscow: Progress Publishers)
[9] Rasskazova V 2019 Teambuilding in innovative startup projects *Scientific review Pedagogical science* **2-3** 66–8 https://science-pedagogy.ru/ru/article/view?id=1898
[10] Ford M 2019 *Rise of the Robots: Technology and the Threat of a Jobless* (Future Alpina non-fiction) p 430
[11] Stevenson M 2011 *An Optimist's Tour of the Future: One Curious Man Sets Out to Answer What's Next?* (Avery No Edition Stated edition) p 384
[12] Krechetov I and Romanenko V 2020 Implementation of adaptive learning methods *The issue of education* **2** 252–77
[13] Kasatkina E, Sairanov A, Nefedov D and Rusyak I 2019 In the Application of genetic algorithms for managing organizational systems in the event of emergency situations *Computer research and modeling* **11** (3) 533–56
[14] Chekanin V, Kovshov E and Hue N 2009 Increase of efficiency of evolutionary algorithms at the decision optimization tasks of objects packing *Control systems and information technologies* **37** (3) 63–7
[15] Koza J 2010 Human-competitive results produced by genetic programming *Genetic Programming and Evolvable Machines* **11** (3-4) 251–84
[16] National science Foundation Press release (National Science Foundation) 2009 *Maybe Robots Dream of Electric Sheep, But Can They Do Science?*
[17] Schmidt M and Lipson H 2009 Distilling Free-Form Natural Laws from Experimental Data *Science* p 324

IOP Conf. Series: Materials Science and Engineering 1027 (2021) 012014 doi:10.1088/1757-899X/1027/1/012014

Hierarchy analysis for monitoring the uniformity of measurements system

I V Krasavin[1], A V Yashin[1] and A K Tulekbaeva[2]

[1] All-Russian Research Institute of Metrological Service, 46 Ozernaya, Moscow 119361, Russia
[2] M. Auezov South Kazakhstan University, Shymkent 160012, Kazakhstan

ilya.krasavin@yandex.ru

Abstract. An approach to monitoring the state of the system for ensuring the uniformity of measurements and forecasting its state, based on the method of hierarchy analysis, is proposed. A decomposition of the system for ensuring the uniformity of measurements is proposed, the further processing of the judgments sequence for decision-making based on paired comparisons, is discussed. The processing of experts judgments in order to obtain a generalized judgment is described.

1. Introduction

One of the main priorities in developing the national system for ensuring the uniformity of measurements is the creation of a mechanism for forecasting the needs of the economy and society in measurements, while the lack of such a mechanism should be attributed to systemic problems.

Monitoring and analysis of trends in priority areas of the economy should provide information about the objects and types of measurements, the required accuracy characteristics.

The development of such forecasts is a complex task that requires appropriate support. To solve this problem the All-Russian Research Institute of Metrology Service in 2019 established the Center for Monitoring and Forecasting. The task of the Center is to develop methods to assess the state of ensuring the uniformity of measurements, to forecast the needs in measurements, to assess the impact of metrology on quality of life and the economy nationwide.

The unity of measurements in the country is reached as a result of functioning of the system of maintenance of uniformity of measurements which is formed by set of legislative and other normative legal acts, normative-technical and methodical documents in the field, participants of system of maintenance of uniformity of measurements, the reference base and park of measuring instruments. Thus, the system for ensuring the uniformity of measurements is a complex ergatic system, and the task of monitoring its state is actual.

The state of the system is assumed as a set of quantitative and qualitative parameters (indicators) characterizing: the structure of the system for ensuring the uniformity of changes; the number of personnel as a whole and by categories; technical equipment of measuring instruments; control, training subsystems, infrastructure of the system for ensuring the uniformity of measurements, scientific support subsystem and regulatory legal basis (figure 1).

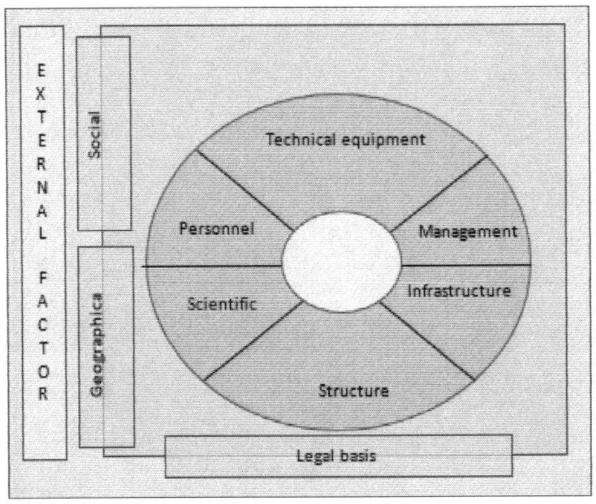

Figure 1. Block diagram of indicators of the state of the system for ensuring the uniformity of measurements.

The quantitative and qualitative composition of the system for ensuring the uniformity of measurements is understood as a set of different for its purpose organizations engaged in activities to ensure the uniformity of measurements.

The structure of the system for ensuring the uniformity of measurements is assumed as a set of hierarchically distributed control bodies for ensuring the uniformity of measurements and the components included in them with the purpose of more effective performance of tasks assigned to them on ensuring the uniformity of measurements.

Evaluation of the state of the system for ensuring the uniformity of measurements is to assess the current state of the system as a whole and its subsystems, the main problem issue is to take into account the integrity of the system, the interaction of subsystems, as well as the presence of indicators, the description of which is possible only at the verbal level.

Thus, the methodological apparatus to assess the state of the system of ensuring the uniformity of measurements, as a complex ergonomic system, should provide an account of interaction between subsystems and the possibility of converting the verbal description of indicators into their numerical values. Nowadays there are numerous methods to solve problems of this type [1, 2]. One of the most effective and relatively simple methods that allow to implement these requirements is the analysis of hierarchies [3–5].

The method of analysis of hierarchies is a systematic procedure for hierarchical representation of elements that determine the essence of a problem. The essence of the method consists in decomposition of the problem into simpler components with further processing of the sequence of judgments of the decision maker by pair comparison. As a result, a relative degree (intensity) of interaction of elements in the hierarchy can be obtained. The method of hierarchy analysis includes procedures for synthesizing multiple judgments, obtaining priority criteria and finding alternative solutions [3, 4].

The main stages of the hierarchy analysis method are the following:

1. Building a qualitative model of the problem in the form of a hierarchy that includes a goal, alternatives to achieve the goal and criteria for assessing the quality of alternatives;
2. Determining the priorities of all elements of the hierarchy using the paired comparison method;
3. Synthesis of global alternative priorities by linear convolution of priorities of elements in the hierarchy;
4. Checking of judgments for consistency.
5. Decision making on the basis of the obtained results.

Let's consider the application of the method of hierarchies analysis to assess the state of the system of ensuring the uniformity of measurements.

The first stage of the method of hierarchy analysis is decomposition of the task of estimating. Thus, the general objective is to evaluate the state of ensuring the uniformity of measurements; it is on the first (highest) level. At the next level, there are goals of ensuring the uniformity of measurements, at the third level, the methods of state regulation in the field of ensuring the uniformity of measurements, and at the fourth level, the subsystems of the system for ensuring the uniformity of measurements (figure 2).

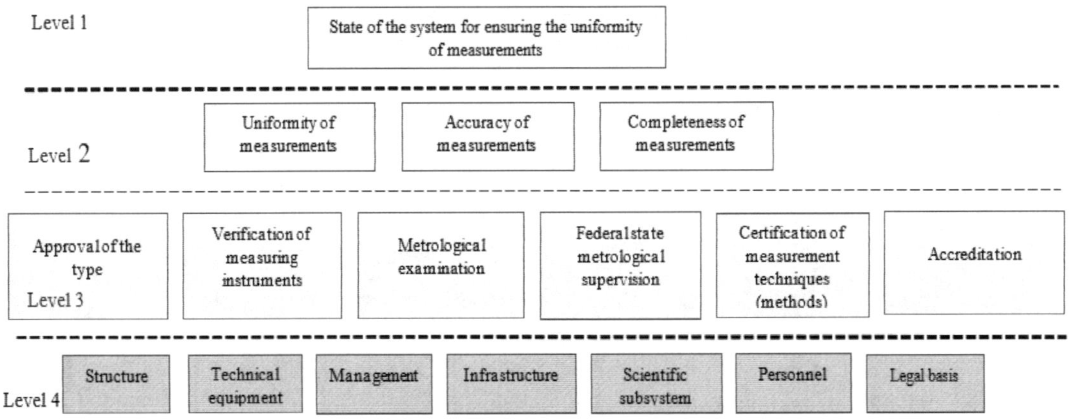

Figure 2. Hierarchical model for assessing the state of the system for ensuring the uniformity of measurements.

In this hierarchy, each element of the level, except for the highest one, depends on all below located elements. The next step establishes the priorities of the criteria, and the evaluation of each of the alternatives by criteria in order to identify the most important of them. The elements are compared in pairs using subjective judgments, numerically evaluated on a scale, with subsequent solving the problem of finding the weight (intensity) of each of the criteria.

To compare the relative importance of the criteria at the second level with respect to the common goal, a matrix of paired comparison of each of the second level alternatives with respect to the first level goal is constructed at the first level. Similar matrices are constructed for paired comparison of each of the alternatives on the third and fourth levels in relation to the criteria of the second and third levels respectively.

The matrix compiled for comparison of relative importance of criteria at the second level for the hierarchy presented in figure 2 is given in table 1.

Table 1. Assessment of the state of the system for ensuring the uniformity of measurements: structure of a matrix of pairwise comparisons.

The state of the system	Uniformity of measurements	The required accuracy of measurements	Completeness of measurements
Uniformity of measurements			
The required accuracy of measurements			
Completeness of measurements			

A structure of matrix of paired comparisons for the Unity of Measurement criteria of the 3rd level of the hierarchy is presented in Table 2, such matrices are compiled for each criterion of the 2nd level, i.e. 3 matrices.

IOP Conf. Series: Materials Science and Engineering 1027 (2021) 012014 doi:10.1088/1757-899X/1027/1/012014

Table 2. Example of a matrix of relative importance for the element of unity of measurements: Matrix of paired comparisons for level 3.

Unity of measurements	Type Approval	Verification of measuring instruments	Metrological examination	Federal State Metrological Supervision	Attestation of measurement methods	Accreditation
Type Approval						
Verification of measuring instruments						
Metrological examination						
Federal State Metrological Supervision						
Attestation of measurement methods						
Accreditation						

Similarly, matrices of pairwise comparisons are formed for each level.

The cells of the given matrices are filled with estimates or judgments about the relative importance of the compared individual subsystems (items) in relation to the goal or criterion indicated at the top. In this case, the matrix cells are planned to be filled with estimates obtained as a result of subjective, but thoughtful judgments of a group of experts who solve this problem. In this case, it is advisable to use the scale for measuring such judgments shown in table 3. The effectiveness of this scale is proved by a large number of applications.

Table 3. Relative importance Scale.

Intensity of relative importance	Definition	Explanations
1	Of equal importance	Equal contribution of two activities to the goal
3	Moderate superiority	Experience and judgment give an easy superiority to one activity over another
5	Significant or strong superiority	Experience and judgment give an considerable superiority to one activity over another
7	A significant advantage	One type of activity is given such a strong superiority that it becomes significant
9	Very strong superiority	The evidence of the superiority of one type of activity over another is most strongly confirmed
2, 4, 6, 8	Intermediate decisions between two adjacent judgments	Apply in a compromise case
Inverse values of the above numbers	If one of the above numbers is obtained when comparing one type of activity with another (for example, 3), then when comparing the second type of activity with the first, we get the inverse value (1/3)	

Let us consider the basic steps of data processing. Obtained in the course of paired comparisons a matrix of paired comparisons of dimension $n \times n$:

$$W = \begin{bmatrix} \frac{w_1}{w_1} & \dots & \frac{w_1}{w_n} \\ \dots & \dots & \dots \\ \frac{w_n}{w_1} & \dots & \frac{w_n}{w_n} \end{bmatrix}. \tag{1}$$

Calculating the vector of local priorities $X = (x_1, x_2, ..., x_n)$:

$$x_i = \frac{\sqrt[n]{\frac{w_i}{w_1} \times \frac{w_i}{w_2} \times \ldots \times \frac{w_i}{w_n}}}{\sum_{i=1}^{n} \sqrt[n]{\frac{w_i}{w_1} \times \frac{w_i}{w_2} \times \ldots \times \frac{w_i}{w_n}}}. \tag{2}$$

The next step is to calculate the own value λ_{max} of matrix W, the IP consistency index and the OS consistency ratio. To control the consistency, the following indicators are calculated:

$$Y = \left(\sum_{j=1}^{n} \frac{w_j}{w_1}, \sum_{j=1}^{n} \frac{w_j}{w_2}, \ldots, \sum_{j=1}^{n} \frac{w_j}{w_n} \right); \tag{3}$$

$$\lambda_{max} = X \cdot Y^T; \tag{4}$$

$$IP = \frac{\lambda_{max} - n}{n-1}; \tag{5}$$

$$OS = \frac{IP}{IP_R}, \tag{6}$$

where IP_R – the normalized consistency index which depends on matrix dimension n (table 4).

Table 4. Standard consistency index.

Matrix size	1	2	3	4	5	6	7	8	9	10
IP_R consistency index	0	0	0.58	0.9	1.12	1.24	1.32	1.41	1.45	1.49

The *OS* value should not exceed 10%. If the *OS* goes beyond these limits, it is necessary to investigate the problem in more detail and check the judgments.

An example of formation of initial data by the expert for the Level 2 is presented in Table 5.

Table 5. Evaluation of the Uniformity of Measurement System Status: Matrix of paired comparisons for Level 2.

State of the Uniformity of Measurement System	Unity of measurements	Required measurement accuracy	Completeness of measurements
Unity of measurements	1.00	3.00	5.00
Required measurement accuracy	0.33	1.00	5.00
Completeness of measurements	0.20	0.20	1.00

As a result of processing the matrix of paired comparisons in accordance with the above-described formulas, we obtain indicators characterizing the degree of influence of the second level elements on the highest level element. An example of the impact assessment results calculated using formula 2 is shown in Figure 3.

Formulas 3–6 calculate the ratio of consistency for the matrix of judgments $OS = 12\%$, which exceeds 10% and indicates that the judgment in the example needs to be reviewed.

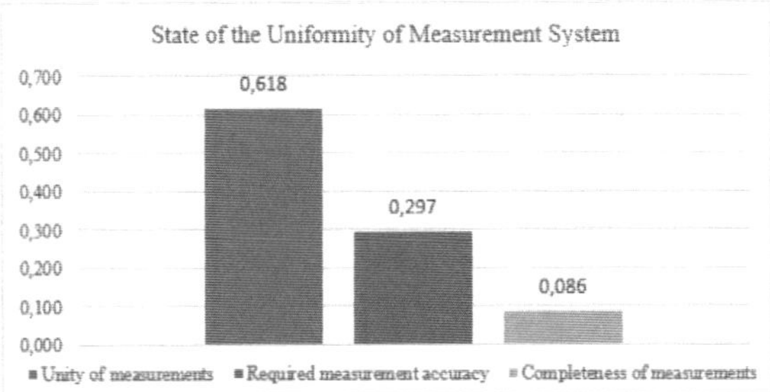

Figure 3. Indicators characterizing the influence of level 2 elements on the Uniformity of Measurement System Status.

An example of filling the matrix of paired comparisons of the third level elements for the second level target «Unity of measurements» is presented in Table 6.

Table 6. Evaluation of the Unity of measurements: Matrix of paired comparisons for Level 3.

Unity of measurements	Type Approval	Verification of measuring instruments	Metrological examination	Federal State Metrological Supervision	Attestation of measurement methods	Accreditation
Type Approval	1.00	7.00	5.00	5.00	3.00	5.00
Verification of measuring instruments	0.14	1.00	7.00	5.00	5.00	5.00
Metrological examination	0.20	0.14	1.00	5.00	3.00	7.00
Federal State Metrological Supervision	0.20	0.20	0.20	1.00	3.00	0.33
Attestation of measurement methods	0.33	0.20	0.33	0.33	1.00	5.00
Accreditation	0.20	0.20	0.14	3.00	0.20	1.00

An example of the influence indicator (2) for the second level target «Unity of measurements» is presented in Figure 4.

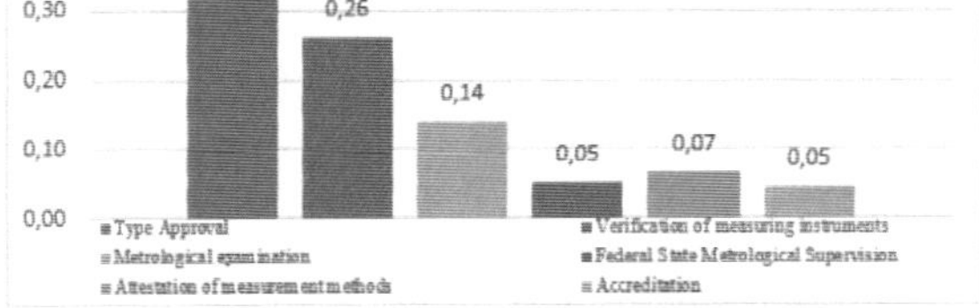

Figure 4. Indicators characterizing the influence of level 3 elements on the Level 2 target «Unity of measurements».

These calculations are performed for each level of the hierarchy, then a global priority vector is calculated. The resulting own vectors (one own value of each criterion) are weighed with the corresponding component of the own vector of the combined criterion, and the results are summed up and normalized.

Thus, application of the method of hierarchy analysis when assessing the state of the system for ensuring uniformity of measurement allows to determine weight (intensity) of influence of this or that subsystem on the system as a whole, which is necessary when determining priority directions of improvement of the metrological system.

References

[1] Wang L et al 2018 *Journal of Computer and Systems Sciences International* **57** (1) 909–14
[2] Diveev A I and Konstantinov S V 2018 *Journal of Computer and Systems Sciences International* **57** (4) 561–80
[3] Saaty T L 2010 *Principia Mathematica Decernendi: Mathematical Principles of Decision Making* (RWS Publications)
[4] Saaty T L and Kearns K P 1985 *Analytical planning. The Organization of systems* (Pergamon press)
[5] Saaty T L 2017 *Cloud of Science* **4** (1) 5–31

IOP Conf. Series: Materials Science and Engineering 1027 (2021) 012015 doi:10.1088/1757-899X/1027/1/012015

Mathematical foundation of system design regulatory controls based on ARP4754A

Vladimir Kulabukhov

Moscow research and production complex "Avionika" named after O.V. Uspenskiy, 7 Obraztsova, Moscow 127055, Russia

kws0704@mail.ru

Abstract. The article provides an interpretation and formal mathematical foundation in the language of modern algebra for *V*-process of designing complex systems recommended by the ARP4754A manual, in the context of a new paradigm in mathematical systems theory developed on the basis of the General principle of isomorphism. This provides a scientific basis for the widely used design approach

1. Introduction

The design problems of modern technical systems are often related to the fact that many of these systems are complex. For complex systems, it is impossible to construct a deterministic model that characterizes the one-to-one relationship between inputs and outputs [1]. Existing models do not allow us to accurately predict the behavior of the system in all possible situations. At the same time, many of the designed systems are critical for their intended purpose to ensure the safety of the operation of the corresponding objects in operating modes. Such systems are called systems with full responsibility. When designing systems with full responsibility, the risks of accidents and catastrophic events should be minimized — such events should become almost impossible. This is especially true for systems related to aviation.

In the absence of deterministic models to ensure a given level of reliability, complex aviation systems must be tested for safety in all possible flight situations under the expected operating conditions after development. However, the complexity of the systems does not allow, firstly, even to anticipate all possible flight situations and, secondly, to perform a very large amount of test work even for obvious situations. In addition to the cost of testing, the time frame for completing tests is also important, which can be very long. The question arises: how to resolve the contradiction between the immensity of the volume, cost and timing of tests on the one hand and ensuring a given level of reliability of equipment and flight safety, on the other hand.

The solution to this contradiction in the design of complex systems was found not in certifying an already developed, ready — made and tested system, but in extending the certification and quality control procedure to the entire development process from the very beginning-from the moment when the requirements for the system and the technical task for its development were formed. In this case, not only the design procedures are subject to verification, but also the qualifications of the personnel involved in their implementation, as well as the structure and level of equipment of the design organization with the necessary equipment and software. In fact, it is assumed that high-quality design procedures performed by qualified personnel under proven conditions guarantee the high quality and reliability of the developed system.

For complex systems, this approach is considered rational. It was tested in practice, proved to be in demand and was fixed in the regulatory documents regulating the procedures for designing and certifying complex systems. One of the main and system-forming in this area is the guide to the development of civil aircraft and systems ARP4754A [2]. The document refers to the full cycle of design and manufacture of aircraft and other complex aircraft systems and recommends using the *V*-model of the design process and its verification, shown in figure 1 [2].

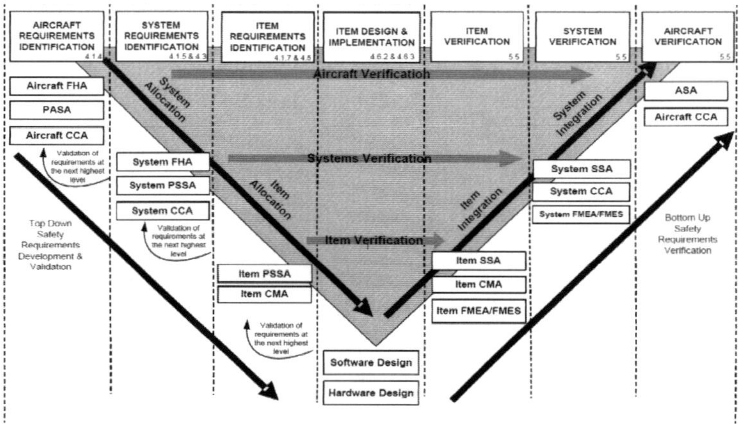

Figure 1. *V*-model of the aviation engineering design process.

The ARP4754A manual addresses the processes of functional safety and aircraft design in the General context of certification. The approach recommended in the guide is essentially a heuristic approach. However, after discussion by the expert community, testing in practice and acceptance for practical guidance by the majority of developers and certifying bodies, the approach has acquired a very high level of objectivity. To date, ARP4754A has been widely recognized and is actually used as a standard for the development and certification of complex aviation systems. The ARP4754A manual is a Central element in the system of documents that govern the full development cycle of the aircraft and its complex systems that are critical to flight safety.

The design approach recommended in ARP4754A is widely used in design practice, developed and discussed in the literature [3–12]. However, questions remain about its scientific validity. This article considers one of the possible variants of such justification in the language of modern mathematics.

In [13, 14], a goal-oriented approach to system design is considered and justified. This approach is similar in many ways to the design approach recommended by ARP4754A: both approaches suggest using a top-down design procedure - from General requirements to specific requirements and further to hardware and software development, to their integration in local devices and in the system as a whole. The difference is that the goal-oriented approach from the very beginning provides for the decomposition of the structure of the designed system into a strictly defined set of independent but interrelated architectures [13, 14]. The interconnection of architectures is determined by the goal of designing a single system that corresponds to a strictly defined purpose, that is, providing for the implementation of specified functions. The sequence of architecture development and their subsequent integration into the system structure as a whole is formalized and prescribed. The approach recommended by ARP4754A does not focus on the structure and architectures of the system – they are chosen by the developer at his discretion, but so that all the requirements and functions prescribed for the system are carefully analyzed and detailed, and then these functions and requirements are implemented both in the local hardware and in the system as a whole. In contrast to the design procedure recommended by ARP4754A, the goal-oriented approach has a fairly strict mathematical justification [14] based on the General principle of isomorphism [1, 15], which is successfully applied to solving various problems in algebraic systems theory [14–17]. We investigate the possibility of applying the General isomorphism principle to formalize the approach to designing complex systems, discussed in the document ARP4754A.

2. Description of the research

Even a brief comparative analysis of the goal-oriented approach [13, 14] and the *V*-process design approach [2] recommended by ARP4754A shows that the *V*-process can also be interpreted and convincingly justified in terms of the General principle of isomorphism in system theory.

In [14] it is shown that from the point of view of mathematics the approaches to top-down design from goals to resources and bottom-up design from resources to goals are equivalent, because there must be a one-to-one correspondence between the objectives (requirements) and resources, based on which the system is implemented. This one-to-one correspondence, from the viewpoint of mathematics, is called isomorphism. The structure of the system is a form of manifestation of this isomorphic mapping of system goals and requirements into resources and, conversely, resources into goals. However, for a number of quality criteria and indicators, such as the level of integration of resources, efficiency and cost of their use in the system structure, it is more preferable to design "top-down" from the goals (requirements) to resources. This design simplifies the formation of the structure itself and its constituent architectures [13, 14]. The formalization of a purposeful approach to system design in the language of algebra is given in [14]. However, the formalization in the language of algebra of the design approach using the *V*-model recommended by ARP4754A will be somewhat different. Let's explain this.

The main result of a new paradigm in algebraic systems theory based on the General principle of isomorphism is *the theorem on the realization* of isomorphism by a composition of maps [1, 15]. Based on figure 2, we present here a statement of the theorem without proof.

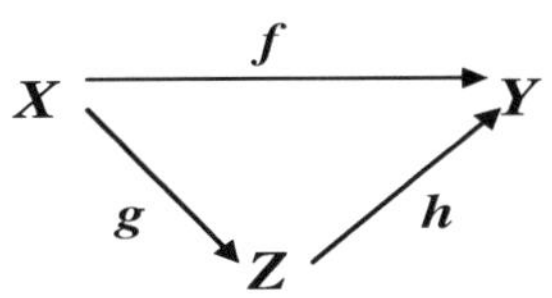

Figure 2. Commutative diagram of the mappings for of the theorem on the realization.

The theorem on the realization: if there is a model $f: X \to Y$ that maps data X to data Y isomorphically and there are mappings $g: X \to Z$ and $h: Z \to Y$ such that the composition $f = hg$ holds, then

a) up to the model-isomorphism *f*, the data *X* and *Y* are accurate and complete;
b) the model *f* is minimal and unique in the sense that if there is another model-an isomorphism $f^*: X \to Y$ that explains the same data, then there is also a unique isomorphic map $\varphi: f^* \to f$ such that the relations $f = \varphi f^*$ and $f^* = \varphi^{-1} f$ are satisfied, that is, the maps *f* and f^* are isomorphic to each other: $f \cong f^*$.
c) the maps h and g in the composition $f = hg$ are isomorphism maps up to the isomorphism of *f*.

The proof of the theorem is given in [1], and the rules and examples of determining inverses for maps appearing in the commutative diagram in Figure 2 are illustrated in [15]. The theorem on the realization essentially repeats the formulation of the General isomorphism principle given in [1]. The main mathematical result contained in the theorem on the realization is that the maps h and g that are not isomorphic separately, outside the commutative diagram, have all the properties of isomorphic maps inside the commutative diagram, that is, they have in this commutative diagram the only inverses h^{-1} and g^{-1} and are thus isomorphisms up to the unconditional isomorphism *f*. The map *f* is inherently isomorphic even outside the commutative diagram and is therefore called an unconditional isomorphism. The composition $f = hg$ may not be unique: there may be another composition

$f = h'g'$ of other maps of h' and g' that implements the same isomorphism *f*. The mappings h' and g' will also satisfy all the conditions of the realization theorem, and these mappings will have their own unique inverse mappings $(h')^{-1}$ and $(g')^{-1}$ inside the corresponding commutative diagram.

From the point of view of our research, it doesn't matter what algebraic structures are meant when we talk about "data" *X* and *Y* in Figure 2. It is important that the map *f* "recalculates" elements and rules for working with them from the algebraic structure *X* to elements and rules for working with them in the algebraic structure *Y*. In terms of this work, "data" *X* will mean the set of goals of the system (requirements for it) and the relationships between them, for example, the relationship of coordination and subordination (subordination) of requirements and goals. "Data" *Y* refers to the set of resources (software, hardware, and others) and the relationships between them that are used to achieve these goals. As examples of relationships on a set of resources, you can specify the relationships of nesting resources and relationships between homogeneous resources (for example, hardware and functional modules are embedded in crates, and crates are combined by link lines in a redundant integrated system). By "data" *Z* we mean some algebraic structure that characterizes the embedding of a set of goals-requirements *X* in a set of local resources *Z* and the rules for working with "local embeddings" in the algebraic structure of *Z*. Then the mapping *g* characterizes the procedure for decomposing global goals *X* (goals-requirements of the highest level) into local goals, that is, building a tree of goals, as well as rules for embedding them in local resources. The mapping *h* interprets the procedure for integrating (combining) local resources *Z*, with local goals nested in them, into a common set of global resources (top-level resources) *Y* that implement system goals as a whole (top-level goals).

Taking into account the mathematical formalism of the theorem on the realization and its meaningful interpretation in terms of our work, the *V*-process of designing and certifying complex systems recommended by the ARP4754A manual can be displayed as nested commutative diagrams shown in figure 3.

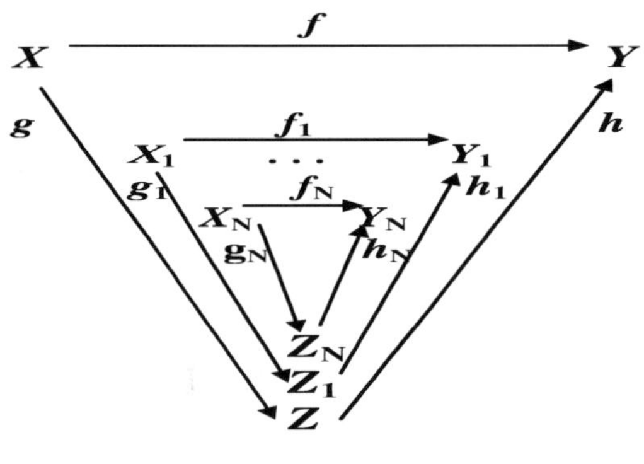

Figure 3. Commutative diagram for the *V*-process.

Figure 3, as well as figure 1, shows how subsets of lower-level local goals $X_1 ... X_N$ are formed on the descending branch of the *V*-process from the initial set of top-level goals *X*-requirements, taking into account the provision of end-to-end traceability of goals on the "top-down"principle. This eliminates the loss of goals, besides, the use of the local mappings $g_1 ... g_N$ provides implementation of each goal in some "own" local resource $Z_1 ... Z_N$. On the ascending branch of the *V*-process, local subsets of resources $Z_1 ... Z_N$ with particular goals implemented in them are integrated into resources of higher and higher levels $Y_1 ... Y_N$ by means of unifying mappings $h_1 ... h_N$. At the same time, each level of consideration of the "goals — resources" mapping must provide its own local isomorphism $f_1 ... f_N$

between goals $X_1...X_N$ and resources $Y_1...Y_N$ of the corresponding level, up to goals-requirements X and resources Y of the highest, system level.

The ARP4754A manual does not describe the actual procedure for decomposing goals-requirements, i.e. it does not contain a procedure for generating local mappings $g_1...g_N$, but it imposes restrictions on the quality of these procedures, provides traceability of goals and requirements when moving to local levels, and contains criteria for evaluating the quality of decomposition. The same applies to procedures $h_1...h_N$ integration of local resources into the General procedure h for creating system resources Y that meet the goals X of creating the system as a whole. The ARP4754A manual requires that at each level of the system review, the composition of local maps h_i and g_i gives a local isomorphism $f_i = h_i g_i$, that is $i = \overline{1...N}$, and provides a one-to-one correspondence between the goals-requirements and resources of the corresponding level.

In addition, one of the main tasks solved by the document ARP4754A in relation to the system design process is the implementation of the superposition principle, which is a consequence of the implementation theorem and the General isomorphism principle [1]. The essence of the superposition principle in the context of this article is that a superposition or, in other words, a certain "sum" (summation will be denoted by a sign $\oplus$) of maps $g_1...g_N$ gave the total mapping $g = g_1 \oplus g_2 \oplus ... \oplus g_N$, "sum" of mappings $h_1...h_N$ gave a General mapping $h = h_1 \oplus h_2 \oplus ... \oplus h_N$, the "sum" of mapping-isomorphisms $f_1...f_N$ gave a General isomorphism $f = f_1 \oplus f_2 \oplus ... \oplus f_N$ between the goals-requirements X and resources Y of the System as a whole.

The implementation of the isomorphism *f*, that is, the creation of a complex system (an aircraft or an aviation system as a whole) cannot be carried out directly and instantly. The isomorphism *f* can only be realized through the implementation of the design process, that is, through the implementation in the form of a composition of processes-maps *h* and *g*, such that $f = hg$. This is the meaning of both the implementation theorem and the General principle of isomorphism [1, 15], and the recommended document ARP4754A *V*-process for designing complex systems. By the theorem on the realization, the maps h and g have the only inverses h^{-1} and g^{-1} in the commutative diagram in figure 3. This means that you can also check the design quality of a complex system in reverse order: from system resources *Y* to local resources *Z* (with local goals implemented in them) and then to system goals-requirements *X*. If another design organization implements an isomorphism *f* between the goals and resources of the system using the composition $f = h'g'$ of other maps h' and g', then these maps will also have to meet the requirements of the theorem on the realization and the *V*-design model recommended by ARP4754A.

Thus, it is shown that the *V*-design process of complex systems, guideline ARP4754A as a deliberate design approach [13, 14], has not only clear the meaning and rationale, but it also allows for a convenient and strictly formalized description in the language of modern algebra and rationale from the point of view of the General principle of isomorphism in mathematical systems theory. This provides a scientific, formal mathematical foundation for the widely used design approach recommended by the ARP4754A document, which ensures the validity of its use in the future.

References

[1] Kulabukhov V S 2018 The General principle of isomorphism in systems theory *Cloud of Science* **5**(3) 400–72
[2] *Guidelines for Development of Civil Aircraft and Systems* 2010 *ARP 4754A* US SAE International: Warrendale, PA, USA, 21 December 2010
[3] Noviello M C, Dimino I, Concilio A, Amoroso F and Pecora R 2019 Aeroelastic assessments and functional hazard analysis of a regional aircraft equipped with morphing winglets *Aerospace* **6** (10) 104 doi: 10.3390/aerospace6100104
[4] LI Xiaoxuna, Zhu Yuanzhenb, Fan Yichenb, Su Duoa 2011 A Comparison of SAE ARP 4754A

and ARP 4754. *The 2nd International Symposium on Aircraft Airworthiness (ISAA 2011) Procedia Engineering* **17** p 400 doi: 10.1016/j.proeng.2011.10.047

[5] Esdras G and Liscouët-Hanke S. 2015 Development of Core Functions for Aircraft Conceptual Design: Methodology and Results *Canadian Aeronautics and Space Institute AERO 2015 Conf.* https://www.researchgate.net/publication/277021129

[6] Mathew P G 2019 Model-Based System Engineering Methodology for Implementing Networked Aircraft Control System on Integrated Modular Avionics — Environmental Control System Case Study https://spectrum.library.concordia.ca/985095/7/GeorgeMathew_MASc_S2019.pdf

[7] *A publication of the International Council on Systems Engineering* December 2019 **22** 4 http://www.panetto.fr/INSIGHT_vol-22-issue-4.pdf

[8] Schulze J 2018 *Architectural Design of a Future Flight Management System Supporting 4D Trajectories. Dissertation* D17. Darmstadt 2018 https://tuprints.ulb.tu-darmstadt.de/8386/1/DissertationSchulze.pdf

[9] Altfeld H H 2010 *Commercial Aircraft Projects: Managing the Development of Highly Complex Products* (Farnham)

[10] *Embedded multi-core systems for mixed criticality applications in dynamic and changeable real-time environments* ARTEMIS Call 2013 project 621429 https://www.artemis-emc2.eu/fileadmin/user_upload/Publications/Deliverables/EMC2_D6.1_SotA_for_System_Qualification_and_Certification_final.pdf

[11] https://afuzion.com/arp4754a-introduction-avionics-systems/

[12] *Certification Specifications and Acceptable Means of Compliance for Large Aeroplanes—CS-25, Amendment 11* 2011 European Aviation Safety Agency: Cologne, Germany

[13] Kulabukhov V S 2015 *Aviakosmicheskoye priborostroyeniye* **12**:11–31

[14] Kulabukhov V 2019 Algebraic Formalization of System Design Based on a Purposeful Approach *IOP Conf. Series: Materials Science and Engineering* **476** 012017 doi: 10.1088/1757-899X/476/1/012017

[15] Kulabukhov V S 2019 A General Principle of Isomorphism: Determining Inverses *Symmetry* **11** 1301 doi:10.3390/sym11101301

[16] Kulabukhov V S 2018 *IOP Conference Series: Materials Science and Engineering* **312** 012016 https://doi.org/10.1088/1757-899X/312/1/012016

[17] Kulabukhov V 2017 *MATEC Web of Conf.* **99** 03008 doi: 10.1051/matecconf/20179903008

IOP Conf. Series: Materials Science and Engineering 1027 (2021) 012016 doi:10.1088/1757-899X/1027/1/012016

To the question of calculating the run distance of the lightweight WIG plan at the stage of preliminary design

A N Luchkov

Moscow Aviation Institute (National Research University), 4 Volokolamskoe shosse, Moscow 125993, Russia

a.luchok.n@gmail.com

Abstract. The work is devoted to the study of the take-off characteristics of high-speed vessels of the WIG type. On the basis of the conducted research, a method for calculating the range of range was developed, considering different types of launch devices during take-off from the water surface. The development of such a method is conditioned by the presence in the design of WIG plans of the problem of determining the most advantageous starting device at takeoff from water under given operating conditions. The proposed method is based on the law of momentum conservation. The method takes into account aerodynamic, hydrodynamic characteristics of WIG plans, characteristics of the power plant from the ship's speed. According to the results of this method it is possible to make a decision on the choice of the most advantageous starting device, providing the performance of the technical task at the stage of preliminary or conceptual design of the ekranoplan. Verification of the technique was carried out using data obtained at the stage of running tests of Orion-10 and Orion-20 WIG plans. The satisfactory coincidence of calculation results and experimental data was obtained.

1. Introduction

Modern equipment is a high-tech product with a life cycle. Cost of life cycle is one of the key characteristics of the product as a whole and includes the cost of development, production, operation and utilization [1]. For passenger ekranoplans the question of reduction of life cycle costs is especial because of high cost uncompetitive in comparison with trains, planes, water transport.

To ensure the specified economic characteristics of the designed lightweight ekranoplan it is necessary, in particular, to determine a sufficient take-off distance taking into account the operating conditions. The take-off characteristics of lightweight WIG plans depend on the aero-hydrodynamic characteristics of the configuration, the ship's power structure and the state of the water area on the approaches to the mooring points (water area of the ekranoplans). Operation of the lightweight WIG plan, which has amphibian quality, implies the possibility of loading and unloading passengers and cargo on the unequipped berths and unloading platforms of the mooring points. Determination of the takeoff characteristics of the lightweight ekranoplans can be carried out in the early stages of design based on the results of testing models, prototypes or numerical simulation.

The analysis [2–10] shows that the established approach to the determination of take-off characteristics of screens is based mainly on the results of experiments with large-scale models in wind tunnels and experimental basins. Calculation of takeoff characteristics of ekranoplans is made mainly by the formulas presented in the sources [4, 5, 8, 11]. Most of the formulas given in the above sources do not take into account on takeoff the force of hydrodynamic resistance of ekranoplans and are almost complete copies of the formulas used in aircraft equipment for runway running [10]. Book [5] provides a more accurate algorithm for calculating the takeoff characteristics of WIG plans, but its use is difficult in view of the application of these formulas' characteristics of power plants, which are usually in closed access.

It is important to note that to solve the problems at the early stages of design, including the determination of the required traction values at takeoff modes, the use of experiments with models to determine the takeoff and landing characteristics [9], seems unreasonably long and expensive.

The results of this study show that at the early stages of design it is possible to use characteristics of takeoff and landing devices, mechanization, hydro-aerodynamic characteristics in order to determine the takeoff and landing distance.

2. Output of the ekranoplan to the cruise mode

It is accepted that the start of the ekranoplan from the water surface under the wave conditions can be considered as a sequence of the following stages [5]:

1 — displacement;
2 — planing;
3 — "ricocheting": the interaction of the bottom with the ridges of waves;
4 — transition to screen cruising movement without contact with water surface.

Figure 1 shows a typical dependence of forces acting on the screen at the start on the stroke speed, including the required and available traction.

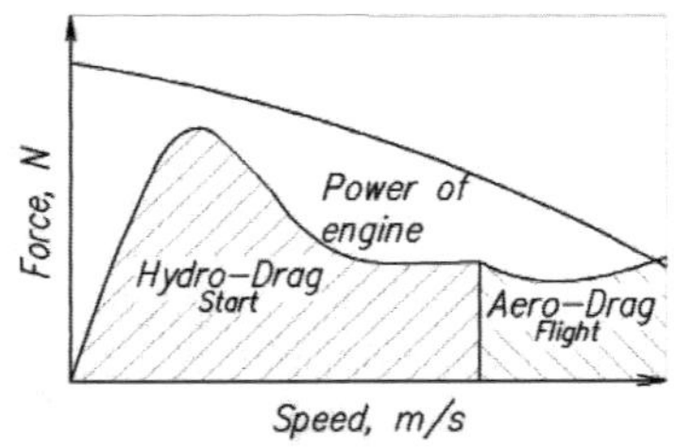

Figure 1. Schedule of forces by the speed of the light WIG plan.

The data of figure 1 qualitatively demonstrate possible ratios of traction forces and resistance at the beginning of movement. At the same time, aerodynamic unloading of the hull of the ekranoplans at the beginning of movement, i.e., removal of the vessel from the aquatic environment into the air, is carried out with the use of energy mechanization in the form of a blow system, a static air cushion and other types of starting devices.

3. Main types of starting devices

There are various ways to provide take-off characteristics of light passenger ekranoplans with take-off weight up to 15 tons:

- Increase of the power plant thrust-to-weight ratio to 0.30–0.35 take-off weight units;
- Application of lifting and marshalling engines in the ship arrangement;
  - Propeller;
  - Turbojet;
- Application of flexible fence and blow molding to create a static and dynamic air cushion.

The choice of the method of take-off provision during the preliminary design must take into account the technical and economic requirements to the vessel as a whole. Repeated examples are known in the history of screen engineering, when high energy consumption of the blowing system leveled all positive technical and economic effect expected from high aerodynamic quality of the wing under the effect of screen effect, and the ship could not withstand the competition in commercial operation. For future projects of ekranoplans, which in many respects will decide the fate of this mode of transport, the development of the layout, which allows to ensure high technical and economic performance, seems fundamental.

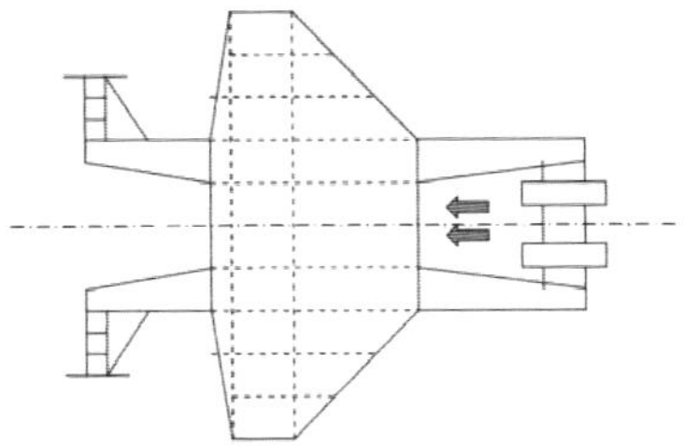

Figure 2. Scheme of the ekranoplan "composite wing" with gas blowing from turbojet engine (view in the plan).

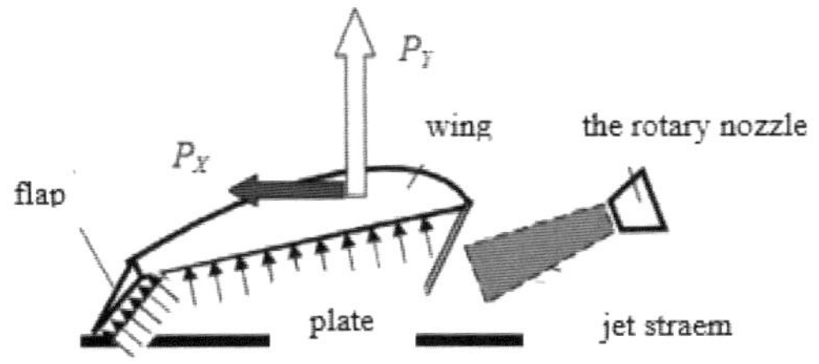

Figure 3. Scheme of gas blowing under the load-bearing wing.

4. Derivation of the formula for calculating the range of the run of the WIG plan

The method for calculating the range of the ekranoplan can be presented in the following stages:

1. Determination of the ekranoplan tear-off speed according to the normative documentation of the project or prototype;
2. Determining the dependencies of the thrust and resistance forces on the speed acting on the device during takeoff, determination of acceleration;
3. Application of kinematic formulas for calculation of range of run.

To begin with, let us derive the formula for calculating the range using Newton's laws and the formula for calculating the range of run through acceleration (1):

$$L = \frac{V_{ga}^2}{2a_a}, \tag{1}$$

L — running distance; V_{ga} — get away speed; a_a — average acceleration, where the average acceleration can be calculated based on Newton's second law and obtain the average:

$$a_a = \frac{P_0(V) - X_{hyd}(V) - X_{aer}(V)}{m_0}, \tag{2}$$

$P_0(V)$ — speed traction control function; $X_{hyd}(V)$ — function of change in hydrodynamic resistance from speed; $X_{aer}(V)$ — function of aerodynamic resistance changes from speed; m_0 — take-off weight.

Expand the subclauses $P_0(V)$, $X_{aer}(V)$:

$$a_a = \frac{P_0\varphi(V)\varphi_в - X_{hyd}(V) - (C_{x0}\frac{\rho V^2 S_\Sigma}{2} + \frac{C_{loc}^2(1-\sigma)}{\pi\lambda_c}\frac{\rho V^2 S_c}{2})}{m_0}, \tag{3}$$

P_0 — total engine thrust; $\varphi(V)$ — speed related thrust function; $\varphi_в(V)$ — function that considers the change in propeller traction by speed; C_{x0} — drag coefficient; ρ — air density; S_Σ — total wing area; C_{loc} — lifting factor of the console; 1-σ — Wieselsberger correction factor; λc — console extension; S_c — console space.

Let's multiply the numerator and denominator for free-fall acceleration g and express equation (4) through design parameters of wing load P_s and traction $\bar{P}$:

$$a_a = g\left(\bar{P}_0\varphi(v)\varphi_в - \bar{X}_{hyd}(V) - C_{x0}\frac{q}{P_s} - \frac{C_{loc}^2(1-\sigma)}{\pi\lambda_c}\frac{q}{P_{sc}}\right), \tag{4}$$

g — free fall acceleration; $\bar{P}_0$ — bench-top traction; $\overline{X_{hyd}}(V)$ — relative hydrodynamic resistance; q — high speed pressure; P_s — wing load; P_{sc} — cantilever load.

IOP Conf. Series: Materials Science and Engineering 1027 (2021) 012016 doi:10.1088/1757-899X/1027/1/012016

One can see that the value of acceleration at any given time depends, among other things, on the speed achieved V. In preliminary calculations, the run length formula can be substituted with an acceleration average value corresponding to the velocity value $V \approx kV_{ga}$. Also, through the average relative value we will express hydrodynamic resistance. With this in mind, we can write the following expressions for calculation of the mean acceleration value at run-up:

$$a_{cp} = g(\bar{P}_0 \varphi(v)\varphi_в - \bar{X}_{гидр_{ср}} - k^2 q\left(C_{x0}\frac{1}{P_s} - \frac{C^2_{ук}(1-\sigma)}{\pi\lambda_к}\frac{1}{P_{зк}} \right), \quad (5)$$

K_{hyd_a} — average hydrodynamic quality.

To determine the coefficient of k and the average relative value of hydrodynamic resistance it is necessary to analyze examples of resistance force dependencies for different systems of unloading the ekranoplan on takeoff. On figures 4, 5 and 6 there are examples of diagrams of change of hydrodynamic resistance on relative speed.

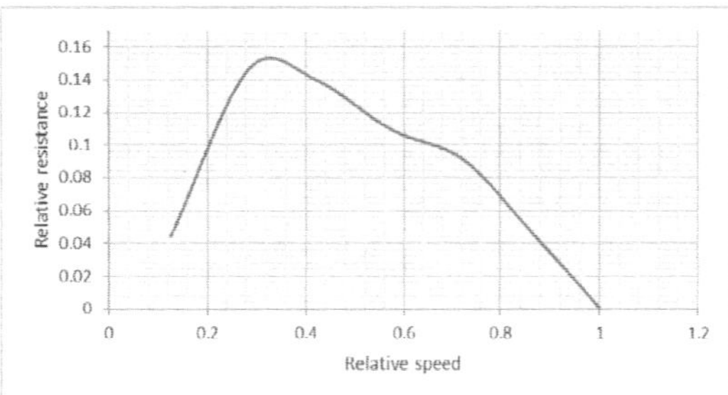

Figure 4. Dependence of the relative resistance force acting on a blowing screen [12].

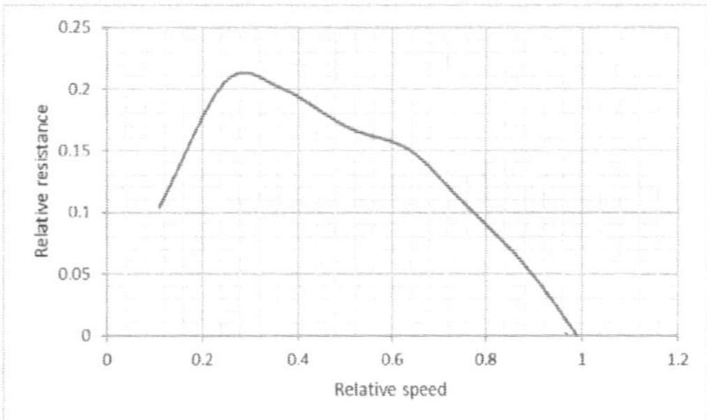

Figure 5. Dependence of the relative resistance force, acting on the screen without starting devices [12].

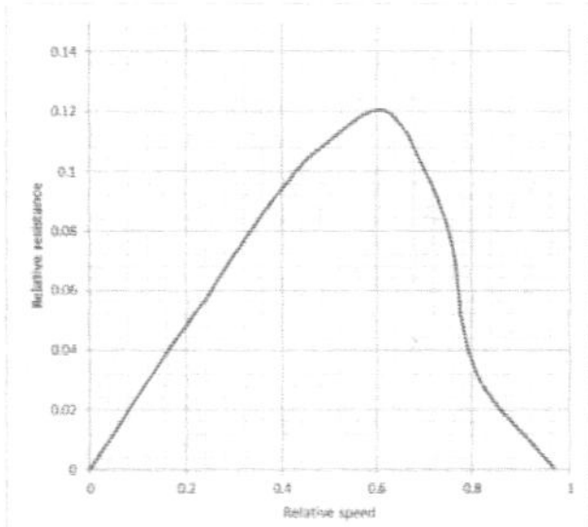

Figure 6. Dependence of the relative resistance force, acting on the screen with the starting device in the form of an airbag [11].

The analysis of the dependencies presented in Figures 4, 5 and 6 shows that the value of average acceleration after overcoming a hump of resistance is reached by a ship at the value of the coefficient of relative speed $k \approx 0.74$ units.

Average relative hydrodynamic resistance can be expressed through average hydrodynamic quality K_{hyd_a}:

$$\bar{X}_{hyd} = \frac{x_{hyd_a}}{mg} = \frac{1}{K_{hyd_a}}. \quad (6)$$

Values of average hydrodynamic quality for ekranoplans with different discharge systems at run-up are given in Table 1.

Table 1. Average hydrodynamic quality. The results of determination of the relative resistance WIG integration function.

	Starting device type		
	No blowing	**With blowing**	**With the starting device in the form of an airbag**
K_{hyd_a}	10.9	8.7	13.5

Average hydrodynamic quality K_{hyd_a} is accepted as corresponding to the hydrodynamic quality at the moment of detachment from the aquatic environment, with the released mechanization of the central plane and deflected hovering ailerons of hypothetical arrangement.

In this regard, let's write down the final equation for calculating the range of WIG plans:

$$L = \frac{V_{ga}^2}{2 * g(\bar{P}_0\varphi(v)\varphi_в - \frac{1}{K_{hyd_a}} - 0,55q\left(C_{x0}\frac{1}{P_s} - \frac{C_{loc}^2(1-\sigma)}{\pi\lambda_c}\frac{1}{P_{sc}}\right)}. \quad (7)$$

5. Checking the output formulas

As a test of the formulas will calculate the take-off range for the WIG plan project Orion-10 with a take-off weight of 4500 kg, equipped with a system of blowing, and WIG plan project Orion-20 with a take-off weight of 14500 kg with the propellers of engines installed on the turntable beam (pylon). The results of testing the head samples of ekranoplans of these projects are presented in Tables 2 and 3.

Table 2. Summary table of Orion-20 ekranoplan characteristics.

Maximum takeoff weight, kg	14500
Power plant type 1	1xturbojet-10B
Power plant type 2	2xM601
Power plant capacity is 1, HP	1040
Power pack capacity 2, HP	750
C_{x0}	0.095
Breaking speed, m/s	47
Range of run, m	3155

Table 3. Summary table of Orion-10 ekranoplan project characteristics.

Maximum takeoff weight, kg	4500
Power plant type	2xcombustion engine
Total thrust of the power plant, kgf	1300
C_x0	0.12
Breaking speed, m/s	42
Range of run, m	1800

As a result of the application of the calculation formula (7), the following characteristics of the range of run are obtained:

IOP Conf. Series: Materials Science and Engineering 1027 (2021) 012016 doi:10.1088/1757-899X/1027/1/012016

Table 4. Calculated ranges of the ekranoplan.

		Difference compared to test results, %
Range of run with blowing Orion-10, m	644	6
Range of run without blowing Orion-20, m	3024	4

6. Conclusion

The technique of estimation of takeoff range of lightweight ekranoplans at early stages of designing is offered. Verification of the method is carried out using the results of Orion-20 and Orion-10 tests. Comparison of the results of distance calculation according to the proposed method with the results of tests shows a good agreement, which confirms the possibility of using the method at the early stages of design of lightweight ekranoplans with take-off weight up to 15 tons.

References

[1] Strelets D Y, Serebryansky S A and Shkurin M V 2019 A digital approach to aircraft product lifecycle management *12-th Int. Conf. Management of large-scale system development (MLSD)*

[2] Belavin N I 1977 *Ekranoplans* (Leningrad: Shipbuilding Publishing House) p 227

[3] Jia Q, Yang W and Yang Z 2016 *Numerical study on aerodynamics of banked wing in ground effect* (Shanghai: Tongji University)

[4] Yun L, Bliault A and Doo J 2010 *WIG Craft and Ekranoplan* (New York: Springer)

[5] Maskalik A I, Kolyzaev B A, Zhukov V I, Radovitsky G L, Sinitsyn D N and Zagorulko L K 2000 *Ekranoplans. Features of Theory and Design* (St. Petersburg: Shipbuilding Publishing House)

[6] Maskalik AI, Nagapetyan R A, Lukyanov A I and others 2005 *Ekranoplans — transport vessels of the XXI century* (St. Petersburg: Shipbuilding Publishing House)

[7] Panchenkov A N, Drachev P T and Lyubimov V I 2006 *Expertise of ekranoplans* (N. Novgorod: VGAVT) p 520

[8] Khimich V L and Chernigin Y P 2011 *Design of Power Plants of WIG-Planes*. (St. Petersburg: Shipbuilding Publishing House) p 496

[9] Sergeyev V G 2018 Conditions of safety and performance of the target task at formation of the ekranoplan appearance at the preliminary design stage *Proc. of the XII Int. Sci. Conf. on Amphibious and Non-aerodrome Av. and Hydroaviation "Hydroaviation Salon-2018"* (Moscow: TsAGI)

[10] Arepiev A N 2006 *Designing of Light Passenger Planes* (Moscow: MAI Publishing House) p 637

[11] Kaljasov P S, Fevralskikh A V and Shabarov V V 2014 Mathematical Modeling of Amphibious Aircraft Aerodynamics on an Air Cushion with Aerodynamic Unloading in the Run Mode on an Air Cushion *Problems of Strength and Plasticity* **76**

[12] http://ekranoplan-ru.narod.ru/C_ANOT.htm

Development of the mathematical model of catalytic oxidation-reduction reaction kinetics for wastewater treatment technology using vortex layer devices

P E Lukashev, M I Sidorov, M E Stavrovskiy, S A Lazarev, A V Ragutkin, E I Korolev and N Sh Gazanova

MIREA - Russian Technological University, 78 Vernadsky Avenue, Moscow 119454, Russia

dyuzya@gmail.com

Abstract. The article considers the mechanisms of intensifying chemical reactions occurring in vortex layer devices. The competition of reagents used as oxidizing agents in water treatment and oxidizing agents appearing during the operation of vortex layer devices is considered. The article offers a description of wastewater treatment process kinetics both for the oxidizing process and the process caused by electrochemical (corrosive) synthesis of highly active reducing agents. The mathematical model is constructed and its analysis is given.

1. Introduction

The goal of the paper [1] was to evaluate the possibility of studying the peculiarities of kinetics in vortex layer devices (VLD) [2, 3] using the mathematical modelling of the "typical" technological process of wastewater treatment. For this purpose, a simple model was considered on the example of catalytic reduction of $Cr(VI)$ compounds with salts of Fe^{2+}. This model demonstrated that the independent additional process of producing highly active oxidizing agents ($\dot{OH}$, H_2O_2 etc.) during the cavitation sonolysis of the treated water can explain the high speed of the purification process in the VLD. As it was noted in [1], the designed simple model has wide possibilities for modification.

In order to set the next similar task of studying the wastewater treatment technological process using mathematical modelling, we need to note the assumption that was not specified in [1]. This assumption consists in the fact that the description of the process actually postulated that the amount of reducing agent (reagent dose) is excessive and always sufficient for the process of oxidizer (contamination) recovery. Therefore, its concentration was not mentioned in the model. However, the reagent dose is an important parameter for the technologist, since an insufficient dose would prevent achieving the required degree of purification, while an excessive amount would turn the reagent into a contaminant. This secondary contamination can pose a problem for the technologist Calculations based on stoichiometry of simple "oxidizer — reducing agent" reaction can only estimate reagent dose. At the same time, the peculiarities of the kinetics of heterogeneous catalytic redox reactions can be unexpected.

2. Mathematical model

First of all, it should be noted that heterogeneous catalysis is associated with the concentration of reagents on the catalyst surface (naturally, this is not its only specifics). Therefore, this fact should be reflected in the mathematical model. Similarly, to the previous study, we will use the simplest models wherever possible. This approach will let us use the results of technological experiments to estimate the model parameters responsible for certain stages of the process and thus, to check the performance of the model by comparing it with the qualitative dynamics of the real object.

Let the following equilibriums of adsorption (desorption) be realized on the surface of the catalyst adsorbent

$$A_N + A_V \underset{k_{-1}}{\overset{k_1}{\rightleftarrows}} A_F; \tag{1.a}$$

$$A_N + B_V \underset{k_{-2}}{\overset{k_2}{\rightleftarrows}} B_F, , \tag{1.b}$$

where A_N, A_F, B_F — surface concentration of active adsorbent centers that are free of adsorbates and occupied by adsorbates: oxidizer and reducing agent, respectively; A_V, B_V — volumetric concentrations of oxidizer and reducing agent; k_1, k_2, k_{-1}, k_{-2} — velocity constants of adsorption and desorption of oxidizer and reducing agent.

From the scheme of reactions (1) follows a system of differential equations of kinetics

$$\frac{dA_F}{dt} = k_1 A_N A_V - k_{-1} A_F; \tag{2.a}$$

$$\frac{dB_F}{dt} = k_2 A_N B_V - k_{-2} B_F. \tag{2.b}$$

Within the quasi-stationary concentration method approach, let us initially assume that the surface area of the catalyst adsorbent is a constant value. This assumption allows us to accept that

$$A_N + A_F + B_F = N_F, \tag{3}$$

i.e. the surface concentration of free and adsorbate occupied active centers of adsorbent-catalyst is a constant value ($N_F = const$). In equilibrium (or for quasi-stationary approximation) $\frac{dA_F}{dt} = \frac{dB_F}{dt} = 0$. From this follows

$$K_{A,ads} = \frac{A_F}{(N_F - A_F - B_F)A_V}; \; K_{B,ads} = \frac{B_F}{(N_F - A_F - B_F)B_V}. \tag{4}$$

Following [1], these expressions can be transformed using the scaled surface concentrations (unit of measurement: $\frac{kg}{m^3}$) instead of surface concentrations (unit of measurement: $\frac{kg}{m^2}$), i.e. instead of (2)–(4) we can write

$$\frac{d\hat{A}_F}{dt} = k_1 \hat{A}_N A_V - k_{-1} \hat{A}_F; \tag{5.a}$$

$$\frac{d\hat{B}_F}{dt} = k_2 \hat{A}_N B_V - k_{-2} \hat{B}_F \tag{5.b}$$

(kinetic constants in (2) and (5) have different dimensions);

$$\hat{A}_N + \hat{A}_F + \hat{B}_F = \hat{N}_F; \tag{6}$$

IOP Conf. Series: Materials Science and Engineering 1027 (2021) 012017 doi:10.1088/1757-899X/1027/1/012017

$$\hat{K}_{A,ads} = \frac{\hat{A}_F}{(\hat{N}_F - \hat{A}_F - \hat{B}_F)A_V}; \; \hat{K}_{B,ads} = \frac{\hat{B}_F}{(\hat{N}_F - \hat{A}_F - \hat{B}_F)B_V}. \quad (7)$$

Let us introduce dimensionless variables $x = \frac{A_V}{A_{V,0}}$; $y = \frac{\hat{A}_F}{A_{V,0}}$; $u = \frac{B_V}{A_{V,0}}$; $v = \frac{\hat{B}_F}{A_{V,0}}$; $N_F^* = \frac{\hat{N}_F}{A_{V,0}}$. After this the equations (7) with dimensionless variables will look as following

$$K^*_{A,ads} = \frac{y}{(N_F^* - y - v)x}; \; K^*_{B,ads} = \frac{v}{(N_F^* - y - v)u}, \quad (8)$$

where $K^*_{A,ads} = \hat{K}_{A,ads}A_{V,0}$ and $K^*_{B,ads} = \hat{K}_{B,ads}A_{V,0}$ — dimensionless variables of adsorption equilibriums. From (8) it is possible to derive equations for dimensionless reduced surface concentrations through volumetric ones

$$y = \frac{K^*_{B,ads}N_F^*u - (K^*_{B,ads}u + 1)v}{K^*_{B,ads}u}; \; y = \frac{K^*_{A,ads}x(N_F^* - v)}{K^*_{A,ads} + 1}; \quad (9)$$

$$v = \frac{K^*_{A,ads}N_F^*x - (K^*_{A,ads}x + 1)y}{K^*_{A,ads}x}; \; v = \frac{K^*_{B,ads}u(N_F^* - y)}{K^*_{A,ads} + 1}. \quad (10)$$

Equations (9) and (10) allow to express the given surface concentrations y and v via volumetric x and u

$$y = N_F^* \frac{K^*_{A,ads}x}{1 + K^*_{A,ads}x + K^*_{B,ads}u}; \; v = N_F^* \frac{K^*_{B,ads}u}{1 + K^*_{A,ads}x + K^*_{B,ads}u}. \quad (11)$$

Removal of the contaminant (oxidizer) A and consumption of the reagent (reducing agent) B is accompanied by the process of catalyst adsorbent production through hydrolysis of the oxidized form of the reducing agent (Fe^{3+}) and to some extent is possible due to hydrolysis of the reduced form of the oxidant (Cr^{3+}). During the hydrolysis of Cr^{3+} colloidal solutions [5]. By analogy with [1] let's write down the equation of kinetics for this process

$$\frac{dG}{dt} = k\hat{A}_F\hat{B}_F G, \quad (12)$$

where $\mu = k\hat{A}_F\hat{B}_F$ — specific production rate of catalytic adsorbent. In dimensionless variables (12) it looks as follows

$$\frac{dG}{dt} = kA_{V,0}^2 yvG, \quad (13)$$

(dimensions $k : \frac{1}{c}\left(\frac{m^3}{kg}\right)^2$); the introduction of dimensionless variables $z = \frac{G}{A_{V,0}}$ and $\tau = \frac{t_0}{kA_{V,0}^2}$ makes it possible to write equation (13) as

$$\frac{dz}{d\tau} = yvz. \quad (14)$$

Besides, the specific reaction rate μ can be defined via dimensionless volumetric concentrations x and u by substituting for it with dimensionless $\mu^* = \frac{\mu}{k}$,

$$\mu = k\frac{N_F^{*2} K_{A,ads}^* K_{B,ads}^* xu}{(1+K_{A,ads}^* x + K_{B,ads}^* u)^2};\ \mu^* = \frac{N_F^{*2} K_{A,ads}^* K_{B,ads}^* xu}{(1+K_{A,ads}^* x + K_{B,ads}^* u)^2}. \quad (15)$$

The equation (15) can be written in a generalized form

$$\mu^* = \frac{\alpha x}{\beta + \gamma x + \delta x^2}, \quad (16.a)$$

where parameters in (16) are defined for a given amount of dimensionless reagent dose u with the equations:

$$\alpha = N_F^{*2} K_{A,ads}^* K_{B,ads}^* u;\ \beta = 1 + 2K_{B,ads}^* u + K_{B,ads}^{*2} u^2;$$
$$\gamma = 2K_{A,ads}^* (1 + K_{B,ads}^* u);\ \delta = K_{A,ads}^{*2}; \quad (16.b)$$

besides, (15) can be presented as

$$\mu^* = \frac{x}{a + bx + cx^2}, \quad (16.a)$$

where $a = \beta/\alpha$, $b = \gamma/\alpha$, $c = \delta/\alpha$ (in this case for constants can be substituted for with three ones, which can be more convenient for evaluations and calculations).

In formulas (16) and (17) parameters $\alpha, \beta, \gamma, \delta$ and a, b, c are the functions of volumetric concentration of oxidizing agent and the thermodynamic affinity of the oxidizing and reducing agents to the adsorptive catalyst surface in the form of equilibrium constants of two-component Langmuir adsorption ($K_{A,ads}^*, K_{B,ads}^*$). The dependence type (17) is illustrated in figure 1 (figure 1a demonstrates the coefficient variation for quadratic terms, figure 1b — for linear terms). These dependencies reflect the fact that the oxidizing and reducing agents compete for the spare active sites of the adsorptive catalyst. If one of the agents prevails, this leads to the reduction of overall oxidation-reduction process speed that results in hydrolysis and the production of adsorptive catalyst. The maximum value of specific speed μ^* is provided by a certain optimal ratio of reagent surface concentrations, which are dependent on their volumetric concentrations.

Formula (17) allows to determine the volumetric concentration of the contaminant component, which corresponds to the maximum specific speed of the treatment process. Analyzing the function $\mu^*(x) = \frac{x}{g(x)}$ to the maximum leads to a simple ration: $x_\mu = \sqrt{\frac{a}{c}}$ (which is derived from equation $(\mu^*(x))'_x = 0$, i.e. $a - cx^2 = 0$).

Thus, for example, for the curve $\alpha(x)$ (figure 1a), which represents $\mu^*(x)$ at $a = 0.05; b = 0.5; c_\alpha = 2.0$ the dimensionless volumetric concentration that corresponds to the maximum $\mu^*(x)$, equals $x_\mu = 0{,}158$.

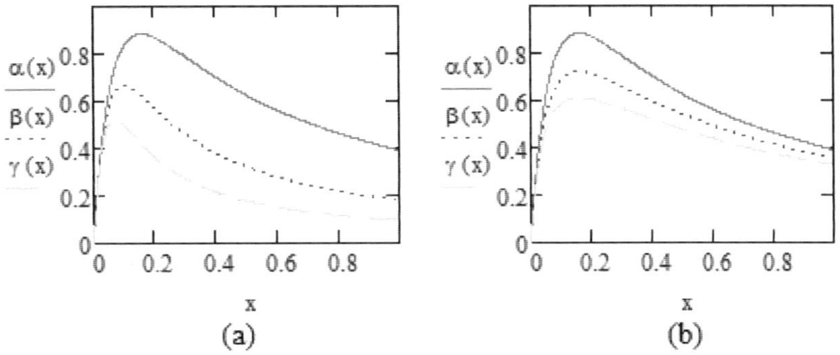

Figure 1. Dependence (17) — specified by $\alpha(x)$, $\beta(x)$, $\gamma(x)$:
(a) — $a = 0.05$; $b = 0.5$; $c_\alpha = 2.0$; $c_\beta = 5.0$; $c_\gamma = 10.0$;
(b) — $a = 0.05$; $b_\alpha = 0.5$; $b_\beta = 0.75$; $b_\gamma = 1.0$; $c = 2.0$.

This illustration demonstrates that the "kinetic" behavior of the system in the case of competitive adsorption significantly differs from the behavior of a system in which the adsorbates do not compete for the spare active sites of the adsorptive catalyst [1]. Such "kinetic" behavior reflects significant peculiarities of the studied system and makes it interesting to perform a more thorough analysis of the kinetics differential equation system for competitive adsorption using numerical methods. In order to do this, we will supplement the system of adsorption equations (1) with a catalytic reaction between reagents (oxidizing and reducing agents) adsorbed on the surface of the catalyst adsorbent. In the result of this reaction the surface is filled with new spare active sites,

$$A_F + B_F \xrightarrow{k_3} fA_N, \tag{18}$$

where f is the stoichiometric coefficient (which is an adjustable parameter in terms of the technological process)

In this case the complete system of kinetic equations (in normalized mass concentrations) looks as following

$$\frac{\mathrm{d}A_V}{\mathrm{d}t} = -k_1 \hat{A}_N A_V + k_{-1} \hat{A}_F; \tag{19.a}$$

$$\frac{\mathrm{d}\hat{A}_F}{\mathrm{d}t} = k_1 \hat{A}_N A_V - k_{-1} \hat{A}_F - k_3 \hat{A}_F \hat{B}_F; \tag{19.b}$$

$$\frac{\mathrm{d}B_V}{\mathrm{d}t} = -k_2 \hat{A}_N \hat{B}_V + k_{-2} \hat{B}_F; \tag{19.c}$$

$$\frac{\mathrm{d}\hat{B}_F}{\mathrm{d}t} = k_2 \hat{A}_N B_V - k_{-2} \hat{B}_F - k_3 \hat{A}_F \hat{B}_F; \tag{19.d}$$

$$\frac{\mathrm{d}\hat{A}_N}{\mathrm{d}t} = -k_1 \hat{A}_N A_V + k_{-1} \hat{A}_F - k_2 \hat{A}_N B_V + k_{-2} \hat{B}_F + f k_3 \hat{A}_F \hat{B}_F. \tag{19.e}$$

The introduction of dimensionless normalized concentration $z = \hat{A}_N / A_{V,0}$ (in addition to the previously introduced x, y, u, v) and time $\tau = t/t_0$ ($t_0 = 1/k_3 A_{V,0}$) allows to write the system (19) as

$$\dot{x} = -\alpha x z + \beta y; \tag{20.a}$$

$$\dot{y} = \alpha xz - \beta y - \varepsilon yv; \tag{20.b}$$

$$\dot{u} = -\gamma uz + \delta v; \tag{20.c}$$

$$\dot{v} = \gamma uz - \delta v - \varepsilon yv; \tag{20.d}$$

$$\hat{z} = -\alpha xz + \beta y - \gamma uz + \delta v + f\varepsilon yv, \tag{20.e}$$

where $\alpha = k_1/k_3$; $\beta = k_{-1}/k_3 A_{V,0}$; $\gamma = k_2/k_3$; $\delta = k_{-2}/k_3 A_{V,0}$; $\varepsilon \approx 1$; f — adjustable (stoichiometric) parameter; $\dot{x} = \mathrm{d}x/\mathrm{d}\tau$; $\dot{y} = \mathrm{d}y/\mathrm{d}\tau$; $\dot{u} = \mathrm{d}u/\mathrm{d}\tau$; $\dot{v} = \mathrm{d}v/\mathrm{d}\tau$; $\dot{z} = \mathrm{d}z/\mathrm{d}\tau$, here α and β are the relative adsorption and desorption rates (i.e. related to the rate of oxidation-reduction reaction).

Preliminary system analysis (20) shows that it has no stationary solution, because if the right parts of equations e.g. (20.a) and (20.b) are equated to zero and then to each other, this leads to the result $(\varepsilon yv = 0)$, which is not assumed by the conditions of the problem. This means the problem has no stationary solution. (1), (18). Therefore, the stationary state of the studied system does not exist in the batch reactor regime, which corresponds with the batch reactor dynamics [5–7].

The results of numerical calculations (MathCad) with varying parameters of the equation system (20) are shown in figure 2–4. Figure 2 demonstrates the situation when the adsorption equilibrium constants of both reagents equal $(\alpha/\beta = \gamma/\delta = 10)$, i.e. they are on par in terms of competing for the surface (for the spare active sites). In the beginning of the process the volumetric concentration of both reagents drastically decrease (figure 2a, c) due to the adsorption on the surface that is free of them (this decrease is not significantly high in absolute figures; figure 2e, f). A similar change is demonstrated by the curves of surface concentrations increase in figure 2 b, d (not shown in larger scale). After this stage the decrease of their volumetric concentration slows down (figure 2e, f), but then the speed of this process increases (figure 2a, c) until the moment when it starts to slow down due to the depletion of reagents in the volume of the reactor. The process completes simultaneously for both reagents.

At first the surface concentration of both reagents increases (figure 2b, d), but as they get depleted in the reactor volume it starts to decrease; however, at almost zero concentration in the volume, the surface concentration does not reach zero. Due to the reaction of formation of new active sites (18), their surface concentration grows along the S-curve (figure 2.g) almost throughout the whole process except for the initial period (figure 2.h) when the concentration reduction is conditioned by the adsorption.

Thus, it can be noted that the kinetics of the studied process in the conditions of competitive adsorption is more complicated than the kinetics of the process when the reagents do not compete for the active surface of the adsorptive catalyst ([1]; figure 2–5).

The comparison of Figures 2 and 3 makes it possible to trace the influence of competition on the studied process. The only difference between the situations illustrated in figure 3 and in figure 2 is that the adsorption equilibrium constant of the second component is 5 times higher than the equilibrium constant of the first component $(\alpha/\beta < \gamma/\delta)$. This leads to higher thermodynamic affinity of the second component with the adsorbent catalyst surface, and its consequent competitive advantage. This results in the fact that the process speed both for the first and for the second component decreases (depletion happens at significantly longer durations — figure 3.a, c).

It is also fixed in the fact that in the initial period the volumetric concentration of the second component decreases at the adsorption stage faster than the volumetric concentration of the first one (figure 3.e, f). Because of this, the surface concentration of the second component increases faster, but it also peaks at later times (also because of the competition with the first component; figure 3.d).

Competition also leads to the fact that the production of new active sites of catalyst adsorption starts to grow dramatically at later stages as well (figure 3.g), as compared to the situation of equal competitive opportunities of both components (figure 2.g).

IOP Conf. Series: Materials Science and Engineering 1027 (2021) 012017 doi:10.1088/1757-899X/1027/1/012017

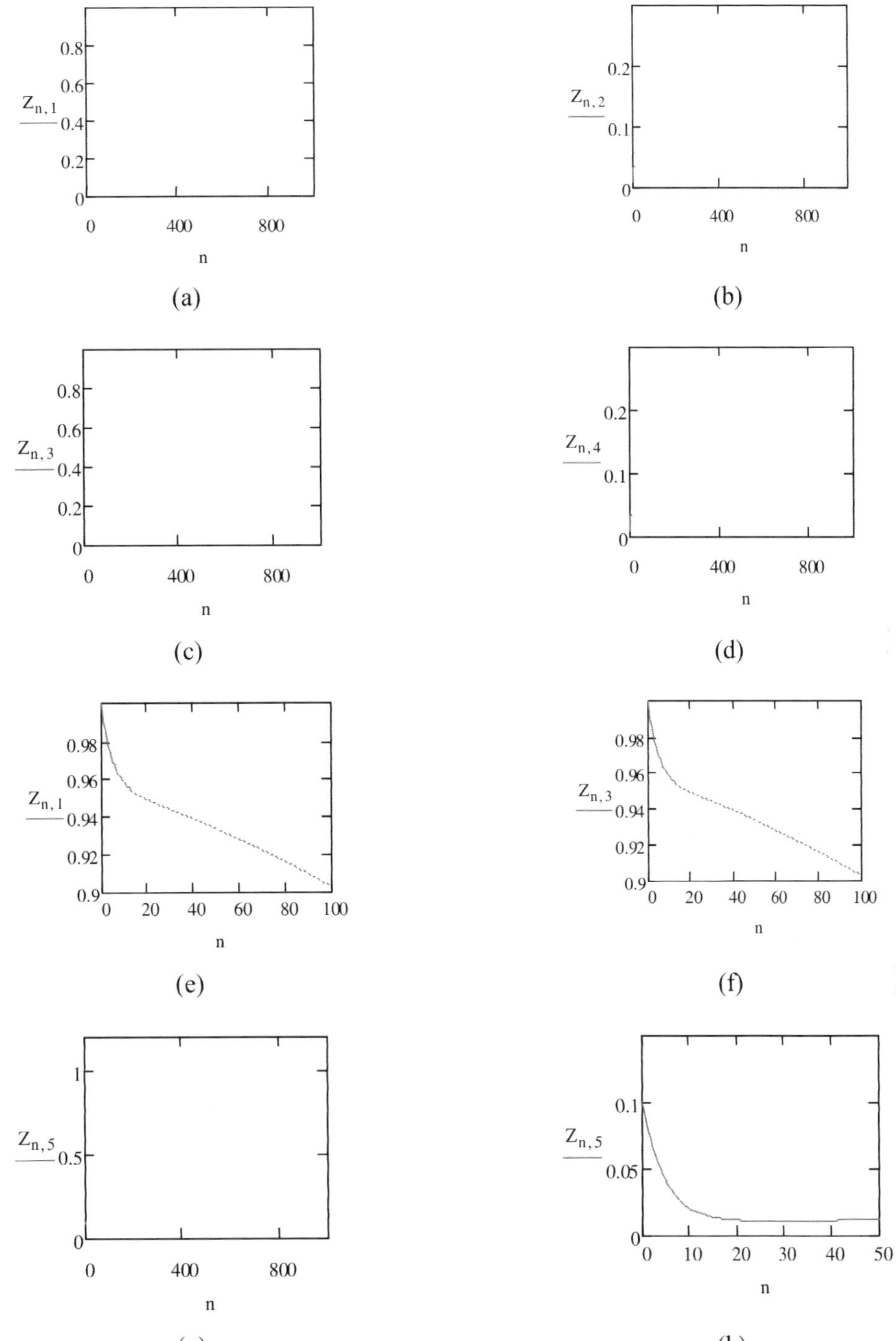

Figure 2. Variation of dimensionless concentrations over time; here $x \to Z_{n,1}$; $y \to Z_{n,2}$; $u \to Z_{n,3}$; $v \to Z_{n,4}$; $z \to Z_{n,5}$; figure (e), (f) and (h) — same as figure (a), (c) and (g), but at the initial period; here $\alpha = 1$; $\beta = 0.1$; $\gamma = 1$; $\delta = 0.1$; $\varepsilon = 1$; $f = 3$; initial conditions: $x_* = 1$; $y_* = 0.01$; $u_* = 1$; $v_* = 0.01$; $z_* = 0.1$.

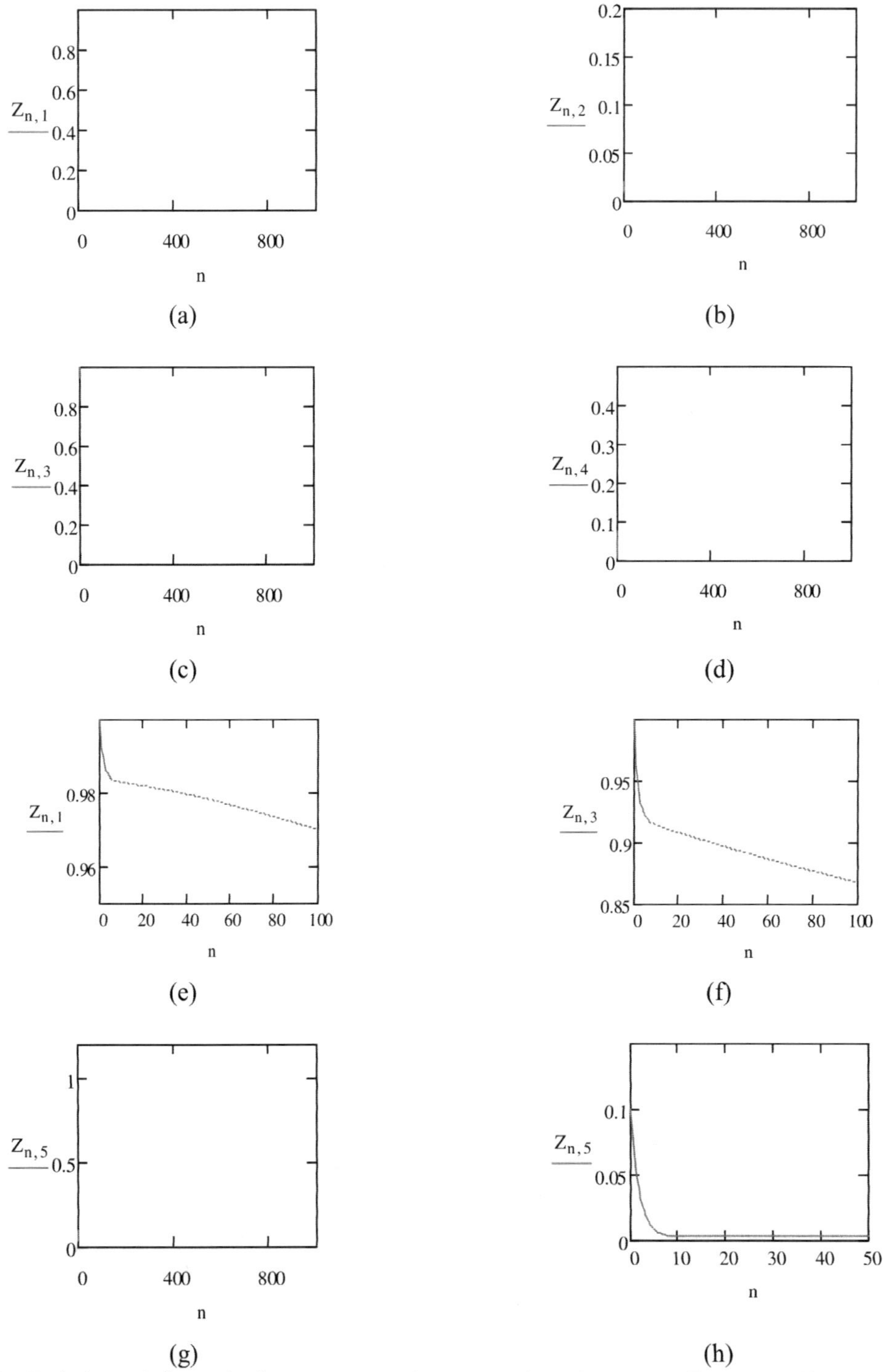

Figure 3. Variation of dimensionless concentrations over time; here $x \to Z_{n,1}$; $y \to Z_{n,2}$; $u \to Z_{n,3}$; $v \to Z_{n,4}$; $z \to Z_{n,5}$; figure (e), (f) and (h) — same as figure (a), (c) and (g), but at the initial period; here $\alpha = 1$; $\beta = 0.1$; $\gamma = 5$; $\delta = 0.1$; $\varepsilon = 1$; $f = 3$; initial conditions: $x_* = 1$; $y_* = 0.01$; $u_* = 1$; $v_* = 0.01$; $z_* = 0.1$.

IOP Conf. Series: Materials Science and Engineering 1027 (2021) 012017 doi:10.1088/1757-899X/1027/1/012017

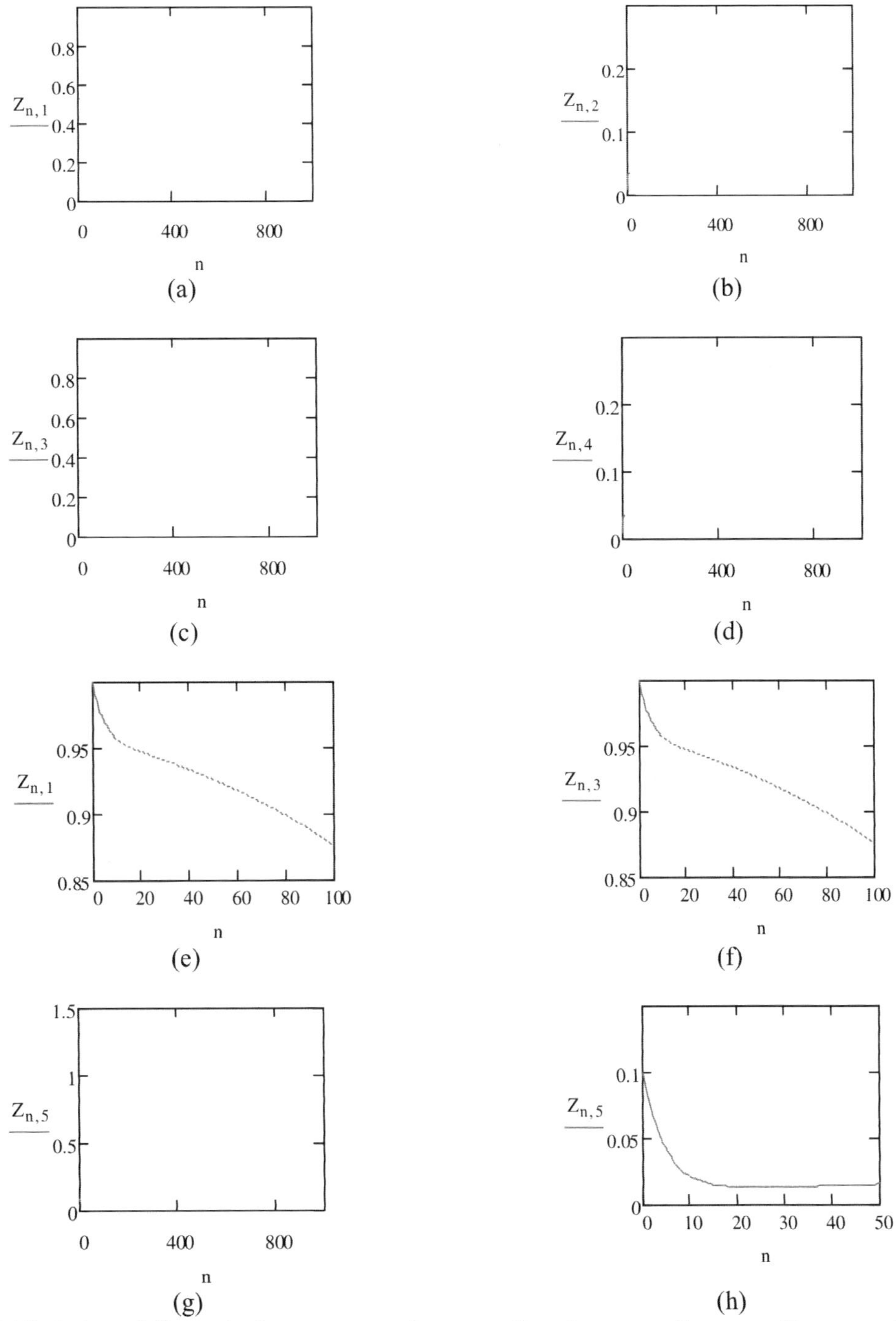

Figure 4. Variation of dimensionless concentrations over time; here $x \to Z_{n,1}$; $y \to Z_{n,2}$; $u \to Z_{n,3}$; $v \to Z_{n,4}$; $z \to Z_{n,5}$; figure (e), (f) and (h) — same as (a), (c) and (g), but at the initial period; here $\alpha = 1$; $\beta = 0.1$; $\gamma = 1$; $\delta = 0.1$; $\varepsilon = 1$; $f = 3$; $g = 0.05$ initial conditions: $x_* = 1$; $y_* = 0.01$; $u_* = 1$; $v_* = 0.01$; $z_* = 0.1$.

Previously, the mathematical model of the kinetics of heterogeneous catalysis of oxidation-reduction reactions showed [1] that an additional independent process that produces highly active oxidizing agents $(\dot{OH}, H_2O_2, O_3)$ and includes cavitation sonolysis reactions [6, 7], which supposedly can take place in VLD [2], leads to a sharp increase in the rate of the main process.

In the mathematical model, this effect is reflected in the modified equation (20.e)

$$\hat{z} = -\alpha xz + \beta y - \gamma uz + \delta v + f\varepsilon yv + gv, \tag{21}$$

where component gv represents the following mechanism: an independent catalytic reaction of oxidizing Fe^{2+} to Fe^{3+}, on the one hand, releases an active site for the second component and on the other hand, produces new active sites through hydrolysis, thus accelerating the total process. This mechanism is confirmed by the results of calculations (figure 4, figure 5). The greatest influence of this independent process is shown in the second case, when the reducing agent occupies the most part of adsorbent-ta-catalyst surface and therefore plays the role of inhibitor (similar to substrate inhibition in enzymatic kinetics [9–11]).

Thus, this variant of demonstrating the contribution of processes initiated in VLD, as well as the previous case [1], shows the possibility of taking into account a number of specific processes, in particular cavitation water sonolysis.

On the other hand, in addition to cavitation and sonolysis, VLD should activate the processes of electrochemical corrosion of the loading particles material ("needles"), especially if they are made of steel [2]. It is known [10] that corrosion processes in the electrolyte medium are strongly activated by friction due to constant removal of electrolysis products (e.g., passive films) as well as due to the elimination of diffusion restrictions on the supply of active solution components (oxidizing agents and reducing agents) to anode and cathode sections of galvanic corrosion pairs. Highly active oxidizing agents formed in sound-chemical reactions can also significantly accelerate this electrochemical process, during which anodic oxidation of metallic iron (Fe^0) may produce Fe^{2+}, which performs the same function as the reagent introduced for cleaning.

In [2] it was also noted that the electromagnetic field that rotates the loaded ferromagnetic particles could also contribute to the activation of electrochemical corrosion by induced electrical charges. In this case, the additional process of corrosion production of active forms of reducing agent (Fe^0, Fe^{2+}) will have an additional increase in speed compared to the rate of tribochemical corrosion. Thus, additional processes conditioned by the cavitation synthesis of highly active oxidizing agents may run in VLD, but also processes conditioned by electrochemical (corrosion) synthesis of highly active reducing agents. This requires a separate qualitative study similar to the one presented above.

The contribution of these additional processes to the total process of water treatment can be reflected within the framework of kinetics models in the form of values of kinetic constants. Therefore, the study makes it possible to conclude that the mathematical models considered in [1] and in this work can be accepted as the basis for the development of the kinetics of wastewater treatment processes that include both sequential and parallel reactions. In this regard, a more in-depth parametric analysis of these models is required.

It should be also noted that chemical and technological systems (CTS) of sewage treatment are usually operated in a continuous mode. This requires mathematical models that correspond to these modes. In its simplest variant this task is solved by the model of continuous stirred-tank reactor (CSTR). The CSTR model may be the most appropriate for the actual process, since in VLD the mixing is very intensive and the retention time is short.

Mathematical models make it is much easier to study the stability of the technological process than the experiment, because the processes that include reactions with complex kinetics often demonstrate unstable behavior. For example, instability is typical for reactions with substrate inhibition (enzyme kinetics) [9–11], which, as noted above, have a description formalism similar to that of reactions involving stages of competitive adsorption. Unstable modes are extremely undesirable for CTS.

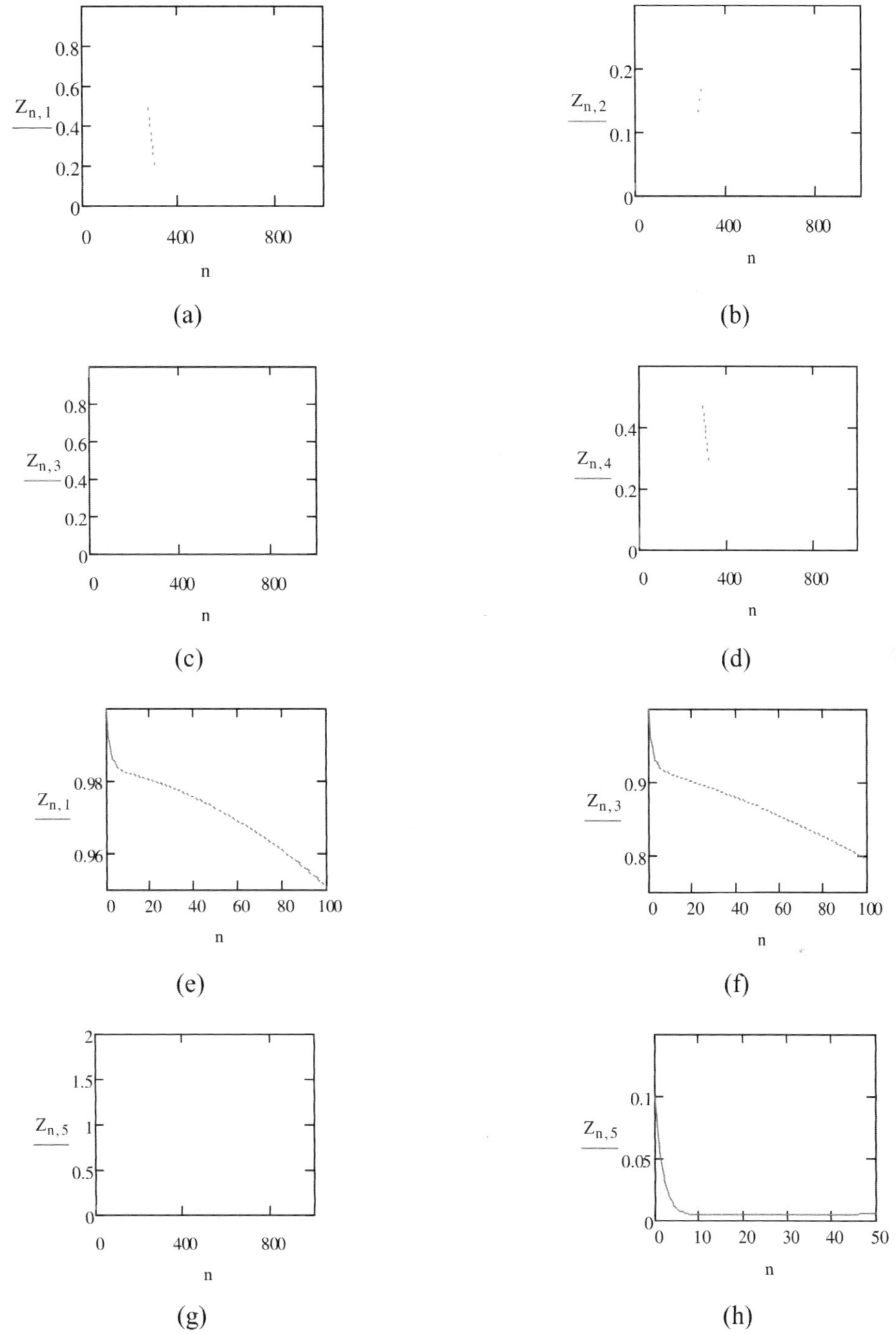

Figure 5. Variation of dimensionless concentrations over time; here $x \to Z_{n,1}$; $y \to Z_{n,2}$; $u \to Z_{n,3}$; $v \to Z_{n,4}$; $z \to Z_{n,5}$; figure (e), (f) and (h) — same as (a), (c) and (g), but at the initial period; here $\alpha = 1$; $\beta = 0.1$; $\gamma = 5$; $\delta = 0.1$; $\varepsilon = 1$; $f = 3$; $g = 0.05$ initial conditions: $x_* = 1$; $y_* = 0.01$; $u_* = 1$; $v_* = 0.01$; $z_* = 0.1$.

All these tasks (within the framework of the general task of XTS mathematical model synthesis) can be studied with the help of the reviewed kinetic model that takes into account both basic (1), (18) and additional reactions, which may require further modifications.

Acknowledgments
This research was supported by Federal State Budget Educational Institution of Higher Education "MIREA – Russian Technological University" (ICMR-8 "Development and Research of Technical Solutions for Purification of Hypertoxic Wastewater").

References

[1] Lazarev S A and Lukashev P E 2020 *The mathematical model of the wastewater treatment process in the vortex layer apparatus*. Manuscript submitted for publication
[2] Adoshev A, Antonov S, Ivashina A and Yastrebov S Investigation of electromagnetic characteristics of ferro-vortex apparatus *Engineering for Rural Development* 2019 780–785
[3] Adoshev A I The peculiarities of calculating the main parameters of the ferro-vortex apparatus *Stavropolye AIC Herald* 2011 1 33–35
[4] Nikolayeva R B *Inorganic chemistry: a study guide in two parts. Part I: Theoretical Foundations of Chemistry* Krasnoyarsk: SFU, 2015
[5] Tovazhnyanskiy L L, Kosheleva M K and Buhkalo S I *General chemical technology: examples, tasks, laboratory work and tests: Study guide* M: INFRA-M, 2015 447 p
[6] Gumerov A M 2014 *Mathematical modeling of chemical and technological processes* Publishing House "Lan" 176 p
[7] Argyros I K and George S *Mathematical Modeling for the Solutions of Equations and Systems of Equations with Applications* Vol 3 Nova Science Publication, New York, 2019
[8] Promtov M A Prospects of using cavitation technologies for the intensification of chemical and technological processes *Tambov State Technical University Herald* 2008 **14** 4
[9] Aponin Y M and Aponina E A The hierarchy of mathematical biology models and numerical-analytical methods of their research (a review) *Mathematical Biology and Bioinformatics* 2007 **2** (2) 347–360
[10] Deng Z, Xia A, Liao Q, Zhu X, Huang Y and Fu Q Laccase pretreatment of wheat straw: effects of the physicochemical characteristics and the kinetics of enzymatic hydrolysis *Biotechnology for biofuels* 2019 **12** (1) 159
[11] Ogurtsov A N *Kinetics of enzymatic reactions* X.: NTU "KhPI", 2007 146 p

IOP Conf. Series: Materials Science and Engineering 1027 (2021) 012018 doi:10.1088/1757-899X/1027/1/012018

Production expert system to determine aircraft take-off and landing performance

A V Melnichuk[1], V A Nesterov[1], V A Sudakov[2,3] and K I Sypalo[4]

[1] Moscow Aviation Institute, 4 Volokolamskoe Shosse, Moscow 125993, Russia
[2] Keldysh Institute of Applied Mathematics, 4 Miusskaya sq., Moscow 125047, Russia
[3] Applied Modelling Science Laboratory, Plekhanov Russian University of Economics, 36 Stremyanny lane, Moscow 117997, Russia
[4] Zhukovsky Central Aerohydrodynamic Institute (TsAGI), 1 Zhukovsky street, Zhukovsky 140180, Moscow Region, Russia

alexander.melnichyuk@gmail.com

Abstract. This paper deals with the design of automated system for take-off and landing performance computing which is applicable to a wide range of aircrafts. The designed solution uses expert systems technology based on rules. The algorithm for logical inference of the expert system is developed.

1. Introduction

Take-off and landing are the critical and the most difficult phases of an aircraft's flight. It is necessary to assess the feasibility of take-off and landing with the actual mass of the aircraft under actual weather conditions, specific runway of a certain airfield during preflight. To do this, it's needed to calculate the take-off and landing aircraft performance. The main purpose of these calculations is to determine maximum take-off and landing weights of an aircraft, speeds for take-off and landing (V_1 – the decision speed, i.e. the maximum speed to which take-off can be safely terminated; V_R – the front landing gear lifting speed; V_2 – the safe take-off speed, V_{REF} – the approach speed).

Take-off and landing performance depends on many factors: aircraft weight, weather conditions at the aerodrome (pressure, temperature, wind speed and direction), runway slope and distances, condition of runway surface, engine thrust on take-off, the braking mode, de-icing system and air conditioning system state, flaps configuration. Take-off and landing characteristics may also be affected by restrictions caused by presence of acceptable defects and deviations of an aircraft (in accordance with MEL – minimum equipment list, and CDL – list of acceptable damages and malfunctions) [1], or defined by operator's policy.

Pilots of Russian aircrafts during pre-flight in most cases use a set of specialized nomograms contained in airplane flight manual (AFM) to calculate take-off and landing performance. Performing calculations in this way "manually" is a complex and time-consuming procedure which requires flight crew increased attention. Complexity of this procedure and lack of time for pre-flight preparation may lead to an error in the calculation. In turn, the use of incorrect results of take-off and landing performance calculations could cause an aviation incident or a plane crash.

Thus, the development of automated pilot's system for take-off and landing performance computing which is applicable to a wide range of aircrafts is an actual scientific and technical task.

IOP Conf. Series: Materials Science and Engineering 1027 (2021) 012018 doi:10.1088/1757-899X/1027/1/012018

Implementation of this system is supposed in the form of a specialized software and hardware complex based on the pilot's tablet computers (EFB — Electronic Flight Bag) [2, 3], which are approved by the aviation authorities [4].

The novelty of the developed solution is in use of expert systems technology, which provides flexibility in the aircraft performance computing by applying rules based on MEL, CDL, AFM and operator policy [5].

2. Production rules for the aircraft performance computing and the logical inference algorithm

Rules which are used to determine aircraft take-off and landing performance can be represented as "if A, then B" implications. To apply such implications, use an expert system that includes a knowledge base for rules storing, a logical inference machine, and a working memory for facts storing are proposed in this paper.

Figure 1 shows an example of the rules of the developed expert system and the process of new facts obtaining during forward logical inference.

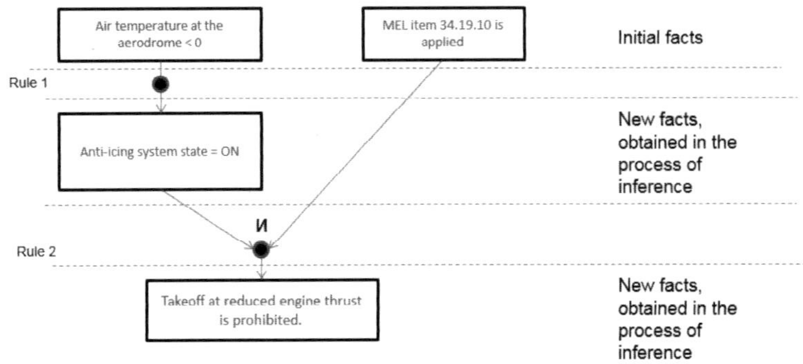

Figure 1. An example of production rules work.

In the process of logical inference, the input data is the initial set of facts and rules in the expert system's knowledge base, the output data is a set of new facts that define new state of working memory. To describe the logical inference algorithm, the following notation is used: WM_0 – initial set of known facts, $f = (v, a)$ – a fact (for example, attribute x with the number v took the value a $(x_v = a)$, v – attribute number, a – attribute value, l – rule number, p_l - priority of a rule l, i – number of the algorithm execution iteration, C – conflict set of rules, CS_i – conflict set of rules at a step i, l^* – number of an activated rule that has the highest priority.

A rule specifies a mapping from a set of given facts X_l at a set of adding facts A_l and removing facts D_l: r_l: $X_l \Rightarrow (A_l, D_l)$, where X_l - a set of facts of antecedents. WM_i is a working memory (initial facts and facts that were obtained by applying of rules at the *i*-th step of the algorithm).

Then the algorithm will take the following form:

1. $i = 0$
2. $i = i + 1$
3. $CS_i = \{ l: (X_l \subseteq WM_{i-1}) \& \left(\forall k \left((0 < k < i) \rightarrow l \notin CS_k\right)\right)\}$
4. If $CS_i = \emptyset$, then finish executing the algorithm.
5. $A = \emptyset$; $D = \emptyset$; $C = CS_i$
6. While $C \neq \emptyset$ execute the cycle:

$$l^* = \arg\max_{l \in C} p_l$$

$$A = A \cup \left[A_{l^*} \setminus \left\{ f_j: \exists f_k \left((f_k \in A) \& (pr_1 f_j = pr_1 f_k)\right)\right\}\right]$$

$$D = D \cup \left[D_{l^*} \setminus \left\{ f_j \colon \exists f_k \left((f_k \in D) \& (pr_1 f_j = pr_1 f_k) \right) \right\} \right]$$
$$C = C \setminus \{l^*\}$$

7. $WM_i = WM_{i-1} \cup A \setminus D$
8. Go to step 2.

The speed of the production expert system (especially when working with large amounts of rules and facts) depends on the inference mechanism. In this paper, the Rete algorithm is used to speed up the inference by minimizing of comparisons in the expert system [6]. In this algorithm, each rule is considered as one or more patterns, each of these is associated with a list of all active elements of working memory which are mapped to it. The input for Rete algorithm are rules of the expert system's knowledge base. As a result, a special acyclic graph is formed, whose nodes correspond to parts of rules conditions.

To implement the expert system and apply the Rete algorithm, it's required to use an appropriate tool. There are many tools for developing of expert systems, and the most common of these are CLIPS, PROLOG, LISP, Drools, and Jess. For tool selecting, a comparative analysis was performed, which resulted in the selecting of CLIPS. Its' advantages are: compatibility with EFB operating systems, open source code, the presence of a built-in inference machine, and the initial purpose for use in the aerospace industry [7], where safety has a critical value.

3. Conclusion

The rules for take-off and landing performance determining can be presented as production rules, which are implemented through inference engine. The time of the expert system work depends on the inference mechanism of the solution. Therefore, to speed up the inference by decreasing number of comparisons in the considered expert system, an approach that allows using the Rete algorithm is proposed. The advantages of CLIPS as well as its use in the aerospace industry where safety is critical (it means that CLIPS system libraries are highly reliable) make it a good choice for crew support expert system implementation.

References

[1] Zontul M 1993 *International Journal of Soft Computing and Engineering* **3** (4) 61–6
[2] Babb T A and Hiers C 2018 *International Journal of Aviation, Aeronautics, and Aerospace* **5** 1–101
[3] Tump R S, van Es G W H, Karwal A K and Verbeek M J 2015 *Best practices for approval of Performance and MB applications on EFBs* (EASA Report)
[4] ICAO Doc 10020 *Manual of electronic Flight Bags (EFB)* 2016
[5] Dutov A V, Sypalo K I, Nesterov V A and Sudakov V A 2018 *Journal of computer and systems sciences international* **57**(2) 230–8
[6] Cirstea H, Kirchner C, Moossen M and Moreau P E 2004 [Electronic resource] Available at: https://hal.inria.fr/file/index/docid/280938/filename/rete.formalisation.pdf
[7] Perfil'ev O V 2014 *Izvestiya Samarskogo nauchnogo centra Rossijskoj akademii nauk* **16** 1(5) 1545–9

IOP Conf. Series: Materials Science and Engineering 1027 (2021) 012019 doi:10.1088/1757-899X/1027/1/012019

Analysis of post buckling state of composite rectangular panel with asymmetric structure on elastic base caused biaxial compression

O Mitrofanov and M Klesareva

Moscow Aviation Institute (National Research University), 4 Volokolamskoe shosse, Moscow 125993, Russia

oleg1mitrofanov@yandex.ru

Abstract. Received analytical solution of geometrically nonlinear problem for investigation of post buckling of composite rectangular panel with asymmetric structure on elastic base. Panel deflection corresponds to conditions of rigidity support. Considered longitudinal compression of panel subject to constraint along the longitudinal edges which corresponds panel loading caused biaxial compression. Results can be used in the analysis of post buckling thin defects such as delaminations in laminated skins.

1. Introduction

According to current regulatory requirements [1] during composite structure design one needs to consider defects which can happen during manufacturing or operation. Analysis of load bearing panels with defect such as delamination proves to be an actual task for many investigators [1–7].

Let us consider a thin rectangular defect such as a delamination in composite panel caused by biaxial compression. Presumably, the specified defect has asymmetric structure and is connected with general panel by adhesive bond. In this case for investigation of behavior of delamination the analytical solution of problem about post buckling behavior of thin panel with asymmetric structure associated with elastic base is considered.

Coupons with dimensions $A*B=150*100$ are normally used for experimental investigation of small defects *(for example, parameters of delaminations can constitute a*b, where a and b can change anywhere from 20 to 40 mm)*. In this case, it is possible to use the standard device which is used for investigation of residual strength of composite materials after impact by ASTM D 7137 (figure 1) [8].

Not to be overlooked here is a specific nature of the coupon fixation which is defined by absence of transversal displacements due to rigid support by lateral clamps. Because of absence of transversal deformations $\varepsilon_y=0$ we will have biaxial compression with transversal compressive force caused by Poisson effect $T_y=T_x c_{yx}/c_y$. Moreover in the considered problem it is advisable to consider initial deflection which can occur and gradually develop under loading of panel with defect.

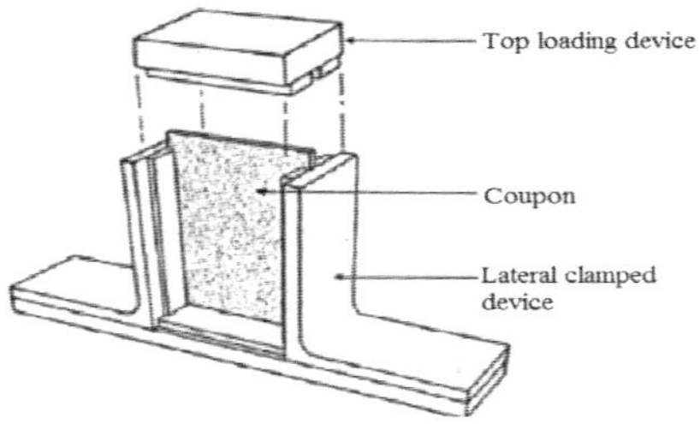

Figure 1. Loading scheme.

2. Basic relationships and problem formulation

For investigation of stress-strain state under post buckling behavior of thin composite panels with asymmetric structure V. Vasilev's theory of composite structures' analysis considering geometrical nonlinearity is employed. Initial relationships are written down for the considered problem taking into account the possibility of initial deflection. Nonlinear strain compatibility equation is written down in the form [9, 10]

$$L_1(F)\text{-}L_2(W)+L_5(W)=0, \tag{1}$$

where

$$L_1(F)=c_y\frac{\partial^4 F}{\partial x^4}+\left(c\text{-}2c_{xy}\right)\frac{\partial^4 F}{\partial x^2\partial y^2}+c_x\frac{\partial^4 F}{\partial y^4},$$

$$L_2(W,W_{нп})=\left(\frac{\partial^2\left(W+W_{нп}\right)}{\partial x\partial y}\right)^2-\left(\frac{\partial^2 W_{нп}}{\partial x\partial y}\right)^2-\frac{\partial^2\left(W+W_{нп}\right)}{\partial x^2}\frac{\partial^2\left(W+W_{нп}\right)}{\partial y^2}+\frac{\partial^2 W_{нп}}{\partial x^2}\frac{\partial^2 W_{нп}}{\partial y^2},$$

$$L_5(W)=B_{yx}\frac{\partial^4 W}{\partial x^4}+\left(B_x+B_y-2C\right)\frac{\partial^4 W}{\partial x^2\partial y^2}+B_{xy}\frac{\partial^4 W}{\partial y^4}.$$

Second nonlinear Karman type equation for composite panel with asymmetric structure with initial deflection is written down in the form

$$L_4(W)-L_3(F,W)-L_6(F)-\sigma_z=0, \tag{2}$$

where

$$L_4(W)=D_x\frac{\partial^4 W}{\partial x^4}+\left(4D+D_{xy}+D_{yx}\right)\frac{\partial^4 W}{\partial x^2\partial y^2}+D_y\frac{\partial^4 W}{\partial y^4},$$

$$L_3(F,W)=\frac{\partial^2 F}{\partial y^2}\frac{\partial^2\left(W+W_{нп}\right)}{\partial x^2}+\frac{\partial^2 F}{\partial x^2}\frac{\partial^2\left(W+W_{нп}\right)}{\partial y^2}-2\frac{\partial^2 F}{\partial x\partial y}\frac{\partial^2\left(W+W_{нп}\right)}{\partial x\partial y},$$

$$L_6(F)=C_{xy}\frac{\partial^4 F}{\partial x^4}+2\left(C_x+C_y-2C\right)\frac{\partial^4 F}{\partial x^2\partial y^2}+C_{yx}\frac{\partial^4 F}{\partial y^4}.$$

where σ_z is elastic base reaction which can be given in the form

$$\sigma_z=C_1W-C_2\left(\frac{\partial^2 W}{\partial x^2}+\frac{\partial^2 W}{\partial y^2}\right), \tag{3}$$

IOP Conf. Series: Materials Science and Engineering 1027 (2021) 012019 doi:10.1088/1757-899X/1027/1/012019

C_i are coefficients of P. Pasternak modulus of subgrade reaction, E_0, μ_0, δ_0 are elastic base parameters (adhesive ply),

$$C_1 = \frac{2E_0}{\delta_0(1-\mu_0^2)}, C_2 = \frac{E_0\delta_0}{12(1-\mu_0)}.$$

3. Applied design techniques

Initial deflection is approximated by a function in the form

$$W_{нп} = f_{нп} \cdot \sin^2 \frac{\pi x}{a} \sin^2 \frac{\pi y}{b} \tag{4}$$

then on initial stage of post buckling under deflection of delamination (when one half-wave is formed in each direction: m=1 and n=1) the function is used that corresponds to comprehensive rigid support

$$W = f \cdot \sin^2 \frac{\pi x}{a} \sin^2 \frac{\pi y}{b}, \tag{5}$$

For analytical solution of the problem Bubnov-Galerkin method is used. Erie function can be determined from relationships

$$N_x = \frac{\partial^2 F}{\partial y^2}, N_y = \frac{\partial^2 F}{\partial x^2}, N_{xy} = -\frac{\partial^2 F}{\partial x \partial y}, \tag{6}$$

Based on equations (1), (4)-(5) in this case after transformation we can get following expression for stress function

$$\begin{aligned} F = &\left(f^2 + 2ff_{нп}\right)\delta\Big\{A_{01}\cos\frac{2\pi x}{a} + A_{02}\cos\frac{2\pi y}{b} + A_{03}\cos\frac{2\pi x}{a}\cos\frac{2\pi y}{b} + \\ &+A_{04}\cos\frac{2\pi x}{a}\cos\frac{4\pi y}{b} + A_{05}\cos\frac{4\pi x}{a}\cos\frac{2\pi y}{b} + A_{06}\cos\frac{4\pi x}{a} + A_{07}\cos\frac{4\pi y}{b}\Big\} + \\ &f\left\{A_{11}\cos\frac{2\pi x}{a} + A_{12}\cos\frac{2\pi y}{b} + A_{13}\cos\frac{2\pi x}{a}\cos\frac{2\pi y}{b}\right\} + \frac{\overline{T}_x y^2}{2} + \frac{\overline{T}_y x^2}{2}, \end{aligned} \tag{7}$$

where

$$A_{01} = \frac{a^2}{32b^2c_y}, A_{02} = \frac{b^2}{32a^2c_x}, A_{03} = -\frac{1}{16\left[c_y\left(\frac{b}{a}\right)^2 + \left(c - 2c_{xy}\right) + c_x\left(\frac{a}{b}\right)^2\right]},$$

$$A_{04} = -\frac{1}{512c_y}\left(\frac{a}{b}\right)^2,$$

$$A_{06} = \frac{1}{32\left[16c_y\left(\frac{b}{a}\right)^2 + 4\left(c - 2c_{xy}\right) + c_x\left(\frac{a}{b}\right)^2\right]}, \tag{8}$$

$$A_{05} = -\frac{1}{512c_x}\left(\frac{b}{a}\right)^2, \quad A_{07} = \frac{1}{32\left[c_y\left(\frac{b}{a}\right)^2 + 4\left(c - 2c_{xy}\right) + 16c_x\left(\frac{a}{b}\right)^2\right]},$$

$$A_{11} = -\frac{B_{yx}}{4c_y}, \quad A_{12} = -\frac{B_{xy}}{4c_x}, \quad A_{13} = \frac{B_{yx}b^4 + B_{xy}a^4 + \left(B_x + B_y + 2C\right)a^2b^2}{16\left(c_y b^4 + \left(c - 2c_{xy}\right)a^2b^2 + c_x a^4\right)}.$$

Next, the membrane stresses which correspond to the initial stage of post buckling behavior can be determined. For example, for longitudinal normal stresses under compression along the X axis considering Poisson effect we have

$$\sigma_x = \frac{1}{\delta}\frac{\partial^2 \mathrm{F}}{\partial y^2} = -\frac{4\pi^2}{\delta b^2}\left(f^2 + 2ff_{нп}\right)\left\{A_{02}\cos\frac{2\pi y}{b} + A_{03}cos\frac{2\pi x}{a}\cos\frac{2\pi y}{b} + \right.$$
$$\left. +4A_{04}\cos\frac{2\pi x}{a}\cos\frac{4\pi y}{b} + A_{05}\cos\frac{4\pi x}{a}\cos\frac{2\pi y}{b} + 4A_{07}\cos\frac{4\pi y}{b}\right\} + \tag{9}$$
$$+\frac{4\pi^2}{\delta b^2}\mathrm{f}\left\{A_{12}\cos\frac{2\pi y}{b} + A_{13}\cos\frac{2\pi x}{a}\cos\frac{2\pi y}{b}\right\} - \frac{T_x}{\delta}\left(1 + \frac{c_{yx}}{c_y}\right).$$

Next as the result of solution of geometrically nonlinear problem by traditional Bubnov–Galerkin method using equilibration and deflection in (3) and (4)

$$\int_0^b\int_0^a [\mathrm{L}_4(\mathrm{W}) \text{-} \mathrm{L}_3(\mathrm{F,W}) \text{ - } \mathrm{L}_6(\mathrm{F}) \text{-} \sigma_z \;]\sin^2\frac{\pi \mathrm{x}}{\mathrm{a}}\sin^2\frac{\pi \mathrm{y}}{\mathrm{b}}\mathrm{dxdy} = 0.$$

after transformation considering the abovementioned Poisson effect the expression for compressive force can be obtained

$$\mathrm{T}_x = \left(\frac{c_y}{c_y + c_{yx}}\right)\left\{\frac{f}{f + f_{нп}}\left[-\frac{3C_1a^2}{\pi^2} + C_2\left(1 + \frac{a^2}{b^2}\right)\right] + \right.$$
$$\left[\frac{4D_x\pi^2}{a^2} + (4D + D_{xy} + D_{yx})\frac{4\pi^2}{3b^2} + \frac{4D_y\pi^2a^2}{b^4}\right] -$$
$$-\frac{4\pi^2 f}{3b^2}\left[\begin{array}{l}2\left(A_{11} + A_{12} - A_{13}\right) + 2\left(f + f_{нп}\right)* \\ *\left(A_{01} + A_{02} - A_{03} + 0.5A_{04} + 0.5A_{05} - A_{06} - A_{07}\right)\end{array}\right] - \tag{10}$$
$$-\frac{16\pi^2 C_{xy} f}{3a^2\left(f + f_{нп}\right)}\left[\left(A_{03} - 2A_{01}\right)\left(f + 2f_{нп}\right) + \left(A_{13} - 2A_{11}\right)\right] -$$
$$-\frac{16\pi^2 C_{yx} a^2 f}{3b^4\left(f + f_{нп}\right)}\left[\left(A_{03} - 2A_{02}\right)\left(f + 2f_{нп}\right) + \left(A_{13} - 2A_{12}\right)\right] -$$
$$\left. -\frac{16\pi^2}{3b^2}\left(C_x + C_y - 2C\right)\frac{f}{f + f_{нп}}\left[A_{13} + fA_{03} + 2f_{нп}A_{03}\right]\right\}$$

IOP Conf. Series: Materials Science and Engineering 1027 (2021) 012019 doi:10.1088/1757-899X/1027/1/012019

To determine the critical forces of delamination of composite laminate of asymmetric structure in linear case the following expression is written down

$$\mathrm{T}_x = \left(\frac{c_y}{c_y + c_{yx}}\right)\left\{\left[-\frac{3C_1 a^2}{\pi^2} + C_2\left(1+\frac{a^2}{b^2}\right)\right] + \right.$$
$$+\left[\frac{4D_x\pi^2}{a^2} + (4D + D_{xy} + D_{yx})\frac{4\pi^2}{3b^2} + \frac{4D_y\pi^2 a^2}{b^4}\right] + \tag{11}$$
$$-\frac{16\pi^2 C_{xy}}{3a^2}\left(A_{13} - 2A_{11}\right) - \frac{16\pi^2 C_{yx} a^2}{3b^4}\left(A_{13} - 2A_{12}\right) -$$
$$\left. -\frac{16\pi^2}{3b^2}\left(C_x + C_y - 2C\right)A_{13}\right\}$$

Except membrane stresses (type (9)) for analysis of geometrically nonlinear behavior of delamination bending stresses of last fibers should be considered which are determined by relationships

$$\sigma_x' = -z[\bar{B}_{11}\frac{\partial^2 W}{\partial x^2} + \bar{B}_{12}\frac{\partial^2 W}{\partial y^2}],\ \sigma_y' = -z[\bar{B}_{12}\frac{\partial^2 W}{\partial x^2} + \bar{B}_{22}\frac{\partial^2 W}{\partial y^2}],\ \tau_{xy}' = -2z\bar{B}_{66}\frac{\partial^2 W}{\partial x \partial y},$$

where $\bar{B}_{mn}$ is given stiffness parameters of composite structure ($B_{mn} = \bar{B}_{mn} \cdot \delta$) [2], $z=\pm\delta/2$, δ is panel thickness.

An example of expression for calculation of longtitudal bending stresses within the considered problem is presented below

$$\sigma_x' = \frac{-z\pi^2}{a^2 b^2}\left[\bar{B}_{11} b^2 \cos\left(\frac{2\pi x}{a}\right)\left(1-\cos\left(\frac{2\pi y}{b}\right)\right) + \bar{B}_{12} a^2 \cos\left(\frac{2\pi y}{b}\right)\left(1-\cos\left(\frac{2\pi x}{a}\right)\right)\right].$$

Thus, when the initial data such as compressive load with considering Poisson effect and given amplitude of initial deflection f_{id} are known, it is possible to calculate the amplitude of deflection f and find stress values in each node.

4. Applied research

Consider the results of calculating the geometrically nonlinear behavior of a thin surface delamination with dimensions a*b=20*20 mm with a stacking sequence of +45°/-45°/0° made from CMU-4 carbon (monolayer thickness equals 0.1 mm). It was assumed that a longitudinal compressive flow $T_x = 10$ kg/mm acts on the composite panel, and a transverse flow T_y=10 kg / mm also occurs. The deflection of the delamination is shown on Figure 2. Normal membrane stress distributions are shown on figures 3, 4.

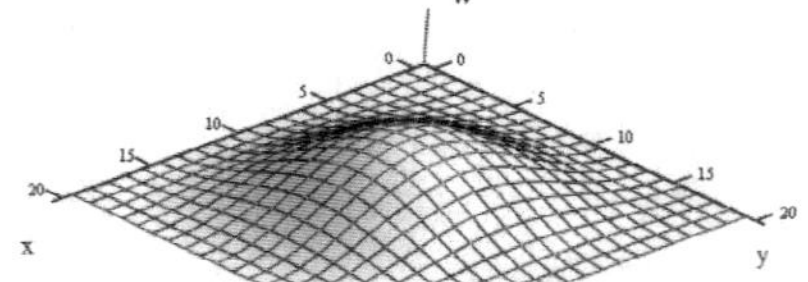

Figure 2. Deflection W.

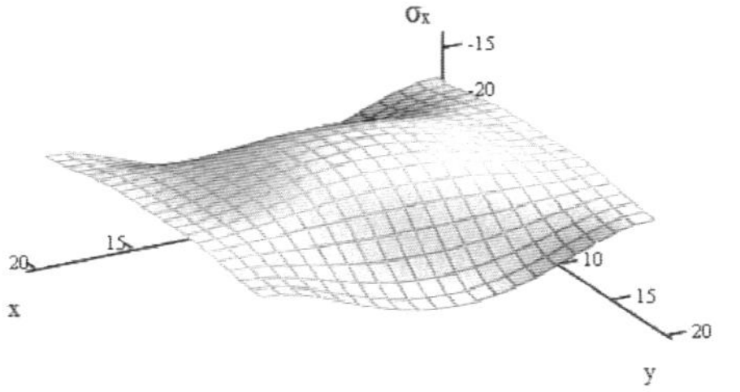

Figure 3. Stress σ.

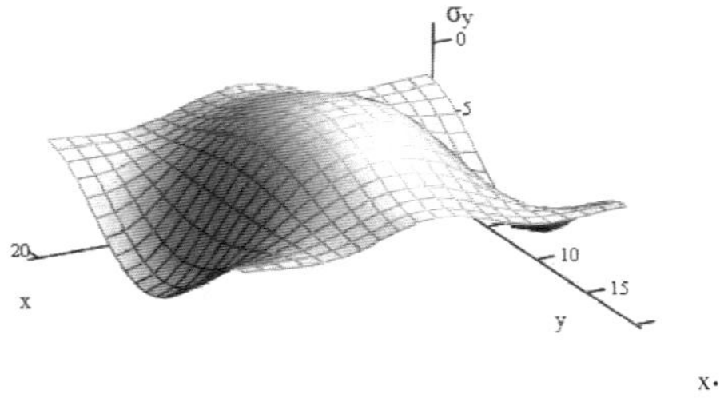

Figure 4. Stress σ_y.

5. Conclusion

In this work a methodology of assessment of post buckling behavior of thin surface delaminations with asymmetric structure which can occur in composite panels is presented.

References

[1] Advisory Circular 20-107A 2009 *Composite Aircraft Structure* 08.09.2009

[2] Bohoeva L A 2007 *Characteristics of strength analysis of structural elements. made from isotropic and composite materials with permissible defects* (Ulan-Ude: ESSUT Publishing House) p 192

[3] Pokrovsky A M and Chermoshentseva A S 2016 Experimantal investigation of aviation materials with defects *Proc. of 15th International Conference «Aviation and cosmonautics»* p 240–2

[4] Riccio A, Raimondo A, Fragale S, Camerlingo F, Gambino B, Toscano C and Tescione D 2014 Delamination buckling and growth phenomena in stiffened composite panels under compression. Part I: An experimental study *J. of Composite Materials* **48** (23) 2843–55

[5] Singh B N, Lal A and Kumar R 2009 Post buckling response of laminated composite plate on elastic foundation with random system properties *J. Communications in Nonlinear Sci. and Numerical Simulation* **14** (1) 284–300

[6] Singh S J and Harsha S P 2019 Exact Solution for Free Vibration and Buckling of Sandwich S-FGM Plates on Pasternak Elastic Foundation with Various Boundary Conditions *Int. J. of Structural Stability and Dynamics* **19** (3) 1950028

[7] Maksimović K, Maksimović M, Vasović-Maksimović I, Rasuo B and Maksimović S 2020 Postbuckling and failure analysis of layered composite panels *FME Transactions* **48** (2) 447–53

[8] ASTM D7137 2017 Standard Test Method for Compressive Residual Strength Properties of Damaged Polymer Matrix Composite Plates, ASTM International, West Conshohocken, PA

[9] Vasilev V V 1988 *Structural mechanics made of composite materials* (Moscow: Mechanical engineering) p 270

[10] Mitrofanov O V 2017 Basic relationships for investigation of post buckling behavior of load bearing panel with asymmetric structure made of composite materials considering initial deflection under *J. Actual problems of modern science* **93** (2) 186–9

Post-buckling behavior estimation of composite panels of asymmetric structure with all-boundary rigid support with shear flows action

O Mitrofanov and M Osman

Moscow Aviation Institute (National Research University), 4 Volokolamskoe shosse, Moscow 125993, Russia

oleg1mitrofanov@yandex.ru

Abstract. Analytical solution of geometrically non-linear problem of post-buckling behavior of thin rectangular plate of asymmetric structure with shear taking rigid support along all sides into account is obtained. Shown solution can be used for post-buckling behavior estimation of surface defect similar to delamination for composite panels.

1. Introduction

Let us examine practical problem of thin surface delamination-type defect stability and post-buckling behavior estimation with shear. It is assumed that defect is of rectangular shape with dimensions. It should be noted that traditional solutions of the problem with hinge support are shown in investigations [3–5]. This investigation takes into account panel rigid support which is more reasonable for delamination behavior analysis. Besides possible composite structure non-symmetry is taken into account [1, 6]. It should also be noted that delamination behavior investigations are of up-to-date nature [7–10].

2. Basic relationships and problem formulation

In order to estimate composite panel mode of deformation of thin local surface defect similar to delamination with shear main correlations of composite structures theory by V. Vasiliev [1] is used. Let us put main correlations for geometrically non-linear problem of composite panel with asymmetric structure down using non-linear equalities [1, 2]. Non-linear equation of strain compatibility is as follows

$$L_1(F)\text{-}L_2(W)+L_5(W)=0, \tag{1}$$

with

$$L_1(F)=c_y\frac{\partial^4 F}{\partial x^4}+\left(c\text{-}2c_{xy}\right)\frac{\partial^4 F}{\partial x^2\partial y^2}+c_x\frac{\partial^4 F}{\partial y^4},\ L_2(W)=\left(\frac{\partial^2 W}{\partial x\partial y}\right)^2-\frac{\partial^2 W}{\partial x^2}\frac{\partial^2 W}{\partial y^2},$$

$$L_5(W)=B_{yx}\frac{\partial^4 W}{\partial x^4}+\left(B_x+B_y-2C\right)\frac{\partial^4 W}{\partial x^2\partial y^2}+B_{xy}\frac{\partial^4 W}{\partial y^4}.$$

with Airy stress function derived from following relationships [1]

$$N_x = \frac{\partial^2 F}{\partial y^2}, N_y = \frac{\partial^2 F}{\partial x^2}, N_{xy} = -\frac{\partial^2 F}{\partial x \partial y} \tag{2}$$

Non-linear von Karman equation is as follows [2]

$$L_4(W) - L_3(F, W) - L_6(F) = 0, \tag{3}$$

with

$$L_4(W) = D_x \frac{\partial^4 W}{\partial x^4} + \left(4D + D_{xy} + D_{yx}\right)\frac{\partial^4 W}{\partial x^2 \partial y^2} + D_y \frac{\partial^4 W}{\partial y^4},$$

$$L_3(F,W) = \frac{\partial^2 F}{\partial y^2}\frac{\partial^2 W}{\partial x^2} + \frac{\partial^2 F}{\partial x^2}\frac{\partial^2 W}{\partial y^2} - 2\frac{\partial^2 F}{\partial x \partial y}\frac{\partial^2 W}{\partial x \partial y},$$

$$L_6(F) = C_{xy}\frac{\partial^4 F}{\partial x^4} + 2\left(C_x + C_y - 2C\right)\frac{\partial^4 F}{\partial x^2 \partial y^2} + C_{yx}\frac{\partial^4 F}{\partial y^4}.$$

Equations (1), (2) set is an initial system of geometrically non-linear problem. Let us use designations shown in the investigation [1]

$$c_x = \frac{B_{22}}{B}, c_y = \frac{B_{11}}{B}, c_{xy} = c_{yx} = \frac{B_{12}}{B}, c = \frac{1}{B_{33}}, \ B = B_{11}B_{22} - B_{12}^2,$$

$$B_x = \frac{1}{B}\left(B_{22}C_{11} - B_{12}C_{12}\right), \ B_y = \frac{1}{B}\left(B_{11}C_{22} - B_{12}C_{12}\right),$$

$$B_{xy} = \frac{1}{B}\left(B_{22}C_{12} - B_{12}C_{22}\right), \ B_{yx} = \frac{1}{B}\left(B_{11}C_{12} - B_{12}C_{11}\right), \ C = \frac{C_{33}}{B_{33}}.$$

$$C_x = c_x\left(C_{11} - \frac{c_{yx}}{c_x}C_{12}\right), \ C_y = c_y\left(C_{22} - \frac{c_{xy}}{c_y}C_{12}\right),$$

$$C_{xy} = c_y\left(C_{12} - \frac{c_{xy}}{c_y}C_{11}\right), \ C_{yx} = c_x\left(C_{12} - \frac{c_{yx}}{c_x}C_{22}\right),$$

$$D_x = D_{11} - C_{11}B_x - C_{12}B_{yx}, \ D_y = D_{22} - C_{22}B_y - C_{12}B_{xy},$$

$$D_{xy} = D_{12} - C_{11}B_{xy} - C_{12}B_y, \ D_{yx} = D_{12} - C_{22}B_{yx} - C_{12}B_x,$$

$$D = \frac{D_{33}B_{33} - C_{33}^2}{B_{33}}.$$

Generalized stiffness of sandwich panel is defined by formulas [1]

$$B_{mn} = I_{mn}^{(0)}, \ C_{mn} = I_{mn}^{(1)} - eI_{mn}^{(0)}, \ D_{mn} = I_{mn}^{(2)} - 2eI_{mn}^{(1)} + e^2 I_{mn}^{(0)},$$

$$I_{mn}^{(r)} = \frac{1}{r+1}\sum_{i=1}^{K} A_{mn}^{(i)}\left(t_i^{r+1} - t_{i-1}^{r+1}\right),\ (r = 0,1,2),\ E_{11,22} = \frac{E_{1,2}}{1-\mu_{12}\mu_{21}},$$

$$A_{11}^{(i)} = \overline{E}_1^{(i)}\cos^4\phi_i + \overline{E}_2^{(i)}\sin^4\phi_i + 2(\overline{E}_1^{(i)}\mu_{12}^{(i)} + 2G_{12}^{(i)})\sin 2\phi_i\cos 2\phi_i,$$

$$A_{12}^{(i)} = A_{21}^{(i)} = \overline{E}_1^{(i)}\mu_{12}^{(i)} + [\overline{E}_1^{(i)} + \overline{E}_2^{(i)} - 2(\overline{E}_1^{(i)}\mu_{12}^{(i)} + 2G_{12}^{(i)})\sin 2\phi_i\cos 2\phi_i],$$

$$A_{22}^{(i)} = \overline{E}_1^{(i)}\sin^4\phi_i + \overline{E}_2^{(i)}\cos^4\phi_i + 2(\overline{E}_1^{(i)}\mu_{12}^{(i)} + 2G_{12}^{(i)})\sin 2\phi_i\cos 2\phi_i,$$

$$A_{33}^{(i)} = [\overline{E}_1^{(i)} + \overline{E}_2^{(i)} - 2\overline{E}_1^{(i)}\mu_{12}^{(i)}]\sin 2\phi_i\cos 2\phi + G_{12}^{(i)}\cos^2 2\phi,$$

with φ_i being reinforcing angle of i-th composite package layer; B_{mn}, C_{mn}, D_{mn} — corresponding membrane, combined and bending stiffness.

3. Applied design methodologies

For considered boundary conditions and panels loading with tangential force deflection function is transformed into

$$\mathrm{W} = \mathrm{f}\cdot\sin^2\frac{\pi\mathrm{y}}{\mathrm{b}}\sin^2\frac{\pi(\mathrm{x}-\alpha\mathrm{y})}{\mathrm{s}}, \tag{4}$$

with α being wave slope ratio, s being distance between node lines.

After deflection (3) substitution into the non-linear equation of strain compatibility (1) stress function taking structure asymmetry into account is derived

$$\begin{aligned} F = f^2\Bigg\{ & A_1\cos\frac{2\pi y}{b} + A_2\cos\frac{4\pi y}{b} + A_3\cos\frac{2\pi(x-\alpha y)}{s} + A_4\cos\frac{4\pi(x-\alpha y)}{s} + \\ & + A_5\cos\frac{2\pi}{bs}\left(bx+sy-\alpha by\right) + A_6\cos\frac{2\pi}{bs}\left(bx-sy-\alpha by\right) + \\ & + A_7\cos\frac{2\pi}{bs}\left(2bx+2sy-\alpha by\right) + A_8\cos\frac{2\pi}{bs}\left(-2bx+2sy+\alpha by\right) + \\ & + A_9\cos\frac{2\pi}{bs}\left(-2bx+sy+2\alpha by\right) + A_{10}\cos\frac{2\pi}{bs}\left(2bx+sy-2\alpha by\right)\Bigg\} + \\ & + f\Bigg\{ A_{11}\cos\frac{2\pi y}{b} + A_{12}\cos\frac{2\pi(x-\alpha y)}{s} + \\ & + A_{13}\cos\frac{2\pi}{bs}\left(bx+sy-\alpha by\right) + A_{14}\cos\frac{2\pi}{bs}\left(bx-sy-\alpha by\right)\Bigg\} + \\ & - \frac{T_x y^2}{2\delta} - \frac{T_y x^2}{2\delta} + \frac{S_{xy}xy}{\delta}, \end{aligned} \tag{5}$$

with following designations introduced:

$$A_1 = \frac{1}{16c_x}\frac{b^2}{s^2},\ A_2 = -\frac{1}{256c_x}\frac{b^2}{s^2},$$

$$A_3 = \frac{s^2}{16b^2}\frac{1}{c_x\alpha^4 + \left(c - 2c_{xy}\right)\alpha^2 + c_y},\quad A_4 = \frac{s^2}{256b^2}\frac{1}{c_x\alpha^4 + \left(c - 2c_{xy}\right)\alpha^2 + c_y},$$

$$A_5 = \frac{s^2b^2}{16}\frac{1}{c_x(s-\alpha b)^4 + \left(c - 2c_{xy}\right)(s-\alpha b)^2 b^2 + c_y b^4},$$

$$A_6 = \frac{s^2b^2}{16}\frac{1}{c_x(s+\alpha b)^4 + \left(c - 2c_{xy}\right)(s+\alpha b)^2 b^2 + c_y b^4},$$

$$A_7 = \frac{s^2b^2}{32}\frac{1}{c_x(2s-\alpha b)^4 + \left(c - 2c_{xy}\right)(2s-\alpha b)^2 b^2 + c_y b^4},$$

$$A_8 = \frac{s^2b^2}{32}\frac{1}{c_x(2s+\alpha b)^4 + \left(c - 2c_{xy}\right)(2s+\alpha b)^2 b^2 + c_y b^4},$$

$$A_9 = \frac{s^2b^2}{32}\frac{1}{c_x(s+2\alpha b)^4 + 4\left(c - 2c_{xy}\right)(s+2\alpha b)^2 b^2 + 16c_y b^4},$$

$$A_{10} = \frac{s^2b^2}{32}\frac{1}{c_x(s-2\alpha b)^4 + 4\left(c - 2c_{xy}\right)(s-2\alpha b)^2 b^2 + 16c_y b^4},$$

$$A_{11} = -\frac{B_{xy}}{4c_x}\frac{1}{s^2},\quad A_{12} = \frac{1}{4}\frac{\left(B_{yx} + \left(B_x + B_y - 2C\right)\alpha^2 + B_{xy}\alpha^4\right)}{c_x\alpha^4 + \left(c - 2c_{xy}\right)\alpha^2 + c_y},$$

$$A_{13} = \frac{1}{8}\frac{\left[B_{yx}b^4 + \left(B_x + B_y - 2C\right)\left(s-\alpha b\right)^2 b^2 + B_{xy}\left(s-\alpha b\right)^4\right]}{c_x(s-\alpha b)^4 + \left(c - 2c_{xy}\right)(s-\alpha b)^2 b^2 + c_y b^4},$$

$$A_{14} = \frac{1}{4}\frac{\left[B_{yx}b^4 + \left(B_x + B_y - 2C\right)\left(s+\alpha b\right)^2 b^2 + B_{xy}\left(s+\alpha b\right)^4\right]}{c_x(s+\alpha b)^4 + \left(c - 2c_{xy}\right)(s+\alpha b)^2 b^2 + c_y b^4},$$

with T_x, T_y, S_{xy} being flows acting on the panel.

Further on using Bubnov-Galerkin procedure for non-linear set (1)–(2) with $f \neq 0$ and tangential force acting following equation is derived

$$\frac{\pi^4}{4b^3s^3}\left\{3D_x b^4 + \left(4D + D_{xy} + D_{yx}\right)\left(3\alpha^2b^2 + s^2\right)b^2 + 3D_y\left(\alpha^2b^2 + s^2\right)^2\right\} +$$

$$+\frac{\pi^4 f^2}{4bs^3}\left[-2A_1 s^2 + 2A_2 s^2 - 3A_3\alpha^2b^2 + 6A_4\alpha^2b^2 + 2A_5\left(s-\alpha b\right)^2 + 2A_6\left(s+\alpha b\right)^2\right] +$$

$$+\frac{\pi^4 f^2}{8bs^3}\left[A_7\left(2s-\alpha b\right)^2 + A_8\left(2s+\alpha b\right)^2 + 2A_9\left(s+2\alpha b\right)^2 + 2A_{10}\left(s-2\alpha b\right)^2\right] +$$

IOP Conf. Series: Materials Science and Engineering 1027 (2021) 012020 doi:10.1088/1757-899X/1027/1/012020

$$+\frac{\pi^4 f}{4bs^3}\left[-2A_{11}s^2-3A_{12}\alpha^2b^2+2A_{13}\left(s-\alpha b\right)^2+2A_{14}\left(s+\alpha b\right)^2\right]+$$
$$+\frac{\pi^4 f^2}{4bs^3}\left[-A_3\left(2\alpha^2b^2+s^2\right)+2A_4\left(3\alpha^2b^2+s^2\right)+2A_5\left(s-\alpha b\right)^2+2A_6\left(s+\alpha b\right)^2\right]+$$
$$+\frac{\pi^4 f^2}{8bs^3}\left[A_7\left(\alpha^2b^2+2\alpha bs+2s^2\right)+5A_8s^2\right]+$$
$$+\frac{\pi^4 f^2}{8bs^3}\left[4A_9\left(2\alpha^2b^2+2\alpha bs+s^2\right)+4A_{10}\left(2\alpha^2b^2-2\alpha bs+s^2\right)\right]+$$
$$+\frac{\pi^4 f}{4bs^3}\left[-A_{12}\left(2\alpha^2b^2+s^2\right)+2A_{13}\left(s-\alpha b\right)^2+2A_{14}\left(s+\alpha b\right)^2\right]+$$
$$-\frac{\pi^4 f^2}{2bs^3}\left[-3A_3\alpha^2b^2+6A_4\alpha^2b^2+A_5\left(s-\alpha b\right)\left(s-2\alpha b\right)+A_6\left(s+\alpha b\right)\left(s+2\alpha b\right)\right]+$$
$$+\frac{\pi^4 f^2}{4bs^3}\left[A_7\left(2s-\alpha b\right)\left(s-\alpha b\right)+A_8\left(2s+\alpha b\right)\left(s+\alpha b\right)+2A_9\left(s+2\alpha b\right)^2+2A_{10}\left(s-2\alpha b\right)^2\right]+$$
$$-\frac{\pi^4 f}{2bs^3}\left[-3A_{12}\alpha^2b^2+A_{13}\left(s-\alpha b\right)\left(s-2\alpha b\right)+A_{14}\left(s+\alpha b\right)\left(s+2\alpha b\right)\right]+$$
$$+\mathrm{f}C_{xy}\frac{\pi^4 b}{s^3}\left(-2A_3+A_5+A_6\right)+\mathrm{f}\left(C_x+C_y-2C\right)\pi^4*\left(\frac{2A_3\alpha^2b}{s^3}+\frac{A_5\left(s-\alpha b\right)^2}{bs^3}+\frac{A_6\left(s+\alpha b\right)^2}{bs^3}\right)+$$
$$+\mathrm{f}C_{yx}\pi^4\left(-\frac{2A_3s}{b^3}-\frac{2A_4\alpha^4b}{s^3}+\frac{A_5\left(s-\alpha b\right)^4}{b^3s^3}+\frac{2A_6\left(s+\alpha b\right)^2}{b^3s^3}\right)+$$
$$+C_{xy}\frac{\pi^4 b}{s^3}\left(-2A_{12}+A_{13}+A_{14}\right)+\left(C_x+C_y-2C\right)\pi^4*\left(\frac{2A_{12}\alpha^2b}{s^3}+\frac{A_{13}\left(s-\alpha b\right)^2}{bs^3}+\frac{A_{14}\left(s+\alpha b\right)^2}{bs^3}\right)+$$
$$+C_{yx}\pi^4\left(-\frac{2A_{12}s}{b^3}+\frac{A_{13}\left(s-\alpha b\right)^2}{b^3s^3}+\frac{2A_{14}\left(s+\alpha b\right)^2}{b^3s^3}\right)=\frac{3\alpha b\pi^2}{8s}\mathrm{S}_{xy}.$$

In order to solve the stability problem equation (5) should be transformed taking initial deflection $(f\rightarrow 0)$ into account. In this case equation (5) is transformed into equality

$$\tau_{xy}=\frac{S_{xy}}{\delta}=\frac{2\pi^2}{3\alpha b^4s^2}\left\{3D_xb^4+\left(4D+D_{xy}+D_{yx}\right)\left(3\alpha^2b^2+s^2\right)b^2+3D_y\left(\alpha^2b^2+s^2\right)^2\right\}+$$
$$+\frac{8}{3\alpha}\frac{\pi^2}{s^2}\left\{C_{xy}\left(-2A_{12}+A_{13}+A_{14}\right)+\left(C_x+C_y-2C\right)\pi^2*\right.$$
$$*\left(2A_{12}\alpha^2+\frac{A_{13}\left(s-\alpha b\right)^2}{bs^3}+\frac{A_{14}\left(s+\alpha b\right)^2}{bs^3}\right)+$$
$$\left.+C_{yx}\pi^4\left(-\frac{2A_{12}s}{b^3}+\frac{A_{13}\left(s-\alpha b\right)^2}{b^3s^3}+\frac{2A_{14}\left(s+\alpha b\right)^2}{b^3s^3}\right)\right\},$$

which contains combined stiffness C_{mn} and takes structure asymmetry into account.

In order to define critical parameters of wave formation following equations shall be used

$$\partial \tau_{xy} / \partial \alpha = 0, \partial \tau_{xy} / \partial s = 0.$$

Shown set allows only numerical solution.

Tangential stress of buckled with shear thin panel is derived from F stress function definition by following formula

$$\tau_{xy} = -\frac{1}{\delta}\frac{\partial^2 F}{\partial x \partial y} = -\frac{S_{xy}}{\delta} - \frac{4f^2\pi^2}{bs^2}\left\{\alpha b A_3 \cos\frac{2\pi(x-\alpha y)}{s} + 4\alpha b A_4 \cos\frac{4\pi(x-\alpha y)}{s} + \right.$$

$$+A_5(\alpha b - s)\cos\frac{2\pi}{bs}(bx+sy-\alpha by) + A_6(\alpha b+s)\cos\frac{2\pi}{bs}(bx-sy-\alpha by)+$$

$$+A_7(\alpha b - 2s)\cos\frac{2\pi}{bs}(2bx+2sy-\alpha by) + A_8(\alpha b+2s)\cos\frac{2\pi}{bs}(-2bx+2sy+\alpha by)+$$

$$\left.+A_9(4\alpha b + 2s)\cos\frac{2\pi}{bs}(-2bx+sy+2\alpha by) + A_{10}(4\alpha b-2s)\cos\frac{2\pi}{bs}(2bx+sy-2\alpha by)\right\}+$$

$$-\frac{4f\pi^2}{bs^2}\left\{\alpha b A_{12}\cos\frac{2\pi(x-\alpha y)}{s} - A_{13}(s-\alpha b)\cos\frac{2\pi}{bs}(bx+sy-\alpha by)+\right.$$

$$\left.+A_{14}(s+\alpha b)\cos\frac{2\pi}{bs}(bx-sy-\alpha by)\right\}.$$

It is noted that derived expression also contains members taking structure asymmetry into account. Besides normal stress can also be defined by similar method.

4. Applied research

For example, consider a thin surface bundle of a CMU-4 carbon fiber panel with a structure of +45°/–45°/0° with a monolayer thickness of δ = 0.1 mm. Let the bundle have geometric dimensions a*b = 80*20 mm and is affected by a tangent flow S_{xy} = 15 kg/m. For figure 1 shows the deflection of a rectangular bundle.

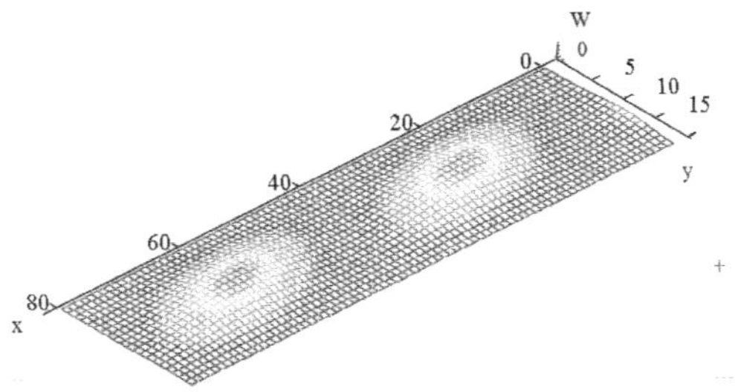

Figure 1. Deflection of the bundle.

5. Conclusion

Obtained analytical solution of geometrically non-linear problem can be used for post-buckling behavior estimation of thin surface rectangular defects similar to asymmetric structure delamination with rigid support along the outline.

References

[1] Vasiliev V 1988 *Composite structures mechanics* (Moscow: Mashinostroenie) p 270
[2] Mitrofanov O 2017 Main correlations for post-buckling behavior investigation of load-bearing

composite panels of asymmetric structure taking initial deflection with compression into account *J. Contemporary science actual problems* **2** 186–9

[3] Balabuhk L 1937 Plywood plates stability *Air fleet technics* **9** 19–38

[4] Azikov N 1993 Investigation of anisotropic plates stability and post-buckling behavior in case of shear *Solid mechanics* **2** 183–9

[5] Grishin V, Dzuba A and Dudarkov J 2013 *Aircraft composite structure elements and joints strength and stability* (Moscow: Phizmatlit) p 270

[6] Karmishin A 1974 Non-homogenous thin-wall elements equations based on minimal stiffness *Applied mechanics* **10** (6) 34–42

[7] Bohoeva L 2007 Peculiarities of isotropic and composite structure elements stress analysis with allowable defects (Ulan-Ude: ESSTU publishing house) p 192

[8] Chermoshentseva A S, Pokrovskiy A M and Bokhoeva L A 2016 The behavior of delaminations in composite materials — experimental results *IOP Conf. Series: Materials Science and Engineering* **116** 012005 doi:10.1088/1757-899X/116/1/012005

[9] Lengvarský P, Bocko J and Hagara M 2016 The Buckling Analysis of the Composite Plates with Different Orientations of Layers *American J. of Mechanical Engineering* **4** (7) 413–7 DOI:10.12691/ajme-4-7-33

[10] Köllner A, Forsbach F and Völlmecke C 2019 Delamination Buckling in Composite Plates: an Analytical Approach to Predict Delamination Growth. *New Achievements in Continuum Mechanics and Thermodynamics (Advanced Structured Materials)* vol 108 ed B Abali, H Altenbach, F dell'Isola, V Eremeyev and A Öchsner (Springer, Cham) https://doi.org/10.1007/978-3-030-13307-8_18

Bayesian recognition of a moving object information-measuring system state: a priori information weight correction

S M Muzhichek[1], G G Sebryakov[1], V I Pavlov[2], S V Tolstych[2] and M B Kenzhekhanova[3]

[1] State Research Institute of Aviation Systems, 7 Viktorenko, Moscow 125319, Russia
[2] Tambov State Technical University, 106 Sovetskay, Tambov 392000, Russia
[3] M. Auezov South Kazakhstan University, Shymkent 160012, Kazakhstan

msm19@yandex.ru

Abstract. Within the framework of the Bayesian approach, a method has been developed for recognizing signal-interference situations that arise during the operation of the information-measuring system of a moving object. The corresponding algorithms are presented. A method for correcting the weight of a priori information in the case of its significant difference from a posteriori information is proposed. One of the possible correction options using the so-called indicators of signals and interference accompanying signs is presented.

1. Introduction

Information-measuring systems are used in the control of moving objects [1]. In this case, the measured components of the moving objects coordinates vector vary over a wide range and can be functionally related. Functionally related are, for example, the range and speed of approach to an external object. The information-measuring system of a moving object is subject to interference, which generates measurement errors. To increase the accuracy of determining moving objects phase coordinates, their information-measuring systems should be developed in the class of systems with a random change in structure in the phase space [2].

Structures are developed in advance, taking into account possible signal-interference situations [3]. To control the change in the structure, it is necessary to recognize the current signal-interference situation, which can be performed in accordance with the Bayesian approach to testing statistical hypotheses [4, 5].

In the paper, within the framework of the Bayesian approach, in addition to measuring the phase coordinates and a priori information about changes in interference situations, indicators of accompanying signs of signals and interference are used [2]. They provide prompt correction of the weight of a priori information in the Bayesian decision-making procedure in cases of its discrepancy with the actual. It is known that if the prior probability of one of the hypotheses is much greater than the prior probabilities of other hypotheses, then with a small number of observations the posterior probability of such a hypothesis will be much higher than the posterior probabilities of other hypotheses. This can lead to erroneous decisions.

2. Formulation of the problem

Let us assume that the vector of phase coordinates of the moving object **x** and the vector of measurement signals **z** in discrete time have the form:

$$\mathbf{x}_{k+1} = \mathbf{f}^{(s)}(\mathbf{x}_k, \mathbf{u}_k, \boldsymbol{\xi}_k); \quad (1)$$

$$\mathbf{z}_k = \mathbf{h}^{(s)}(\mathbf{x}_k, \boldsymbol{\zeta}_k), \quad (2)$$

where **u** is the vector of controls; ξ, ζ are the object and measurement noise vectors, respectively; **f**, **h** — known vector functions of vector arguments; s is the number of the structure (state) of the moving object and its information-measuring system, $s = \overline{1,S}$, S is the number of structures, k is the index of the discrete moment in time.

Based on the results of the preliminary analysis, the following is determined: a list of interferences; a list of possible signal and noise situations that make up a complete group of incompatible events; a priori intensities of the transition $\tilde{\gamma}^{(ls)}$ from the s-th to the l-th and $\tilde{\gamma}^{(sl)}$from the l-th to the s-th signal-noise situations; accompanying signs of interference impact on the information-measuring system. To detect accompanying signs, the output signals of the corresponding channels of the information-measuring system are used.

The process of changing a concomitant feature is a degenerate conditional Markov chain having two states (Pr = 1 or Pr = 0), which is determined for fixed values of s_k and x_k by the probability:

$$q_n(Pr, k | s_k^T, x_k^T, k), \quad (3)$$

where $n = \overline{1,N}$ — feature number; N — is the number of features; Pr — is the value of the feature indicator, $Pr = \overline{0,1}$; $s_k^T = |s_{1,k}, \dots, S_k|$ — is the vector of the set of scalar indices of possible interference situations $s = \overline{1,S}$; $x_k^T = |x_{1,k}, \dots, x_{i,k}|$ — is the vector of the set of scalar values of the i-th phase coordinate of the moving object, $i = \overline{1,I}$, I is the number of measured phase coordinates. This is a degenerate case of a Markov chain, when the probability of transition to the state Pr = 0 or to the state Pr = 1 is equal to the steady-state value of the probability of this state. In the case when the probability of the appearance of a sign of interference does not depend on the phase coordinate, expression (3), called the input indicator function J, takes the form:

$$J_{n,k} = q_n(Pr, k | s_k^T, k), \; n = \overline{1,N}, \; Pr = \overline{0,1}, \; s = \overline{1,S}. \quad (4)$$

The functioning of the feature indicator is described by the conditional probability of transition from r_{k-1} to r_k state [3]:

$$\pi_n(J, Pr) = \pi_n(r_k, k | J_{n,k-1}, Pr, r_{k-1}, k-1), n = \overline{1,N}, r_k = \overline{0,1}. \quad (5)$$

Obviously, the 0 and 1 states of the indicator of the Pr feature correspond to the absence and presence of the accompanying noise feature.

If the interference leads to the appearance of several accompanying signs, then the probability π (J, Pr) of the presence of the interference according to the valuess of the indicators can be represented as:

$$\pi(J, Pr) = \prod_{n=1}^{N} \pi_n(J, Pr), \; n = \overline{1,N}, \; Pr = \overline{0,1}. \quad (6)$$

It is required, based on (1)–(6) and accounting for $\tilde{\gamma}^{(ls)}, \tilde{\gamma}^{(sl)}$, to determine the current signal-interference situation and the corresponding structure number of the information-measuring system of the moving object. It is also required to propose a rational option for correcting the weight of a priori information in cases of its inconsistency with the actual one.

3. Recognition of signal-interference situations

Recognition of the signal-interference situation that has developed at the current time can be performed in accordance with the Bayes formula:

$$\hat{P}_k^{(s)} = \left[\frac{\tilde{P}^{(s)} P(Pr, z|s)}{\sum_{s=1}^{S} \tilde{P}^{(s)} P(Pr, z|s)}\right]_k, \tag{7}$$

where $\widehat{P}^{(s)}$, $\widetilde{P}^{(s)}$ are the a posteriori and a priori probabilities of the s-th signal-interference situation, respectively; $P(\mathrm{Pr}, z|s)$ is the probability of a joint change in the states of indicators of accompanying signs of signals and interferences of a moving object when the s-th signal-interference situation occurs at the k-th moment.

In cases where there is no information about the predominant effect on the information-measuring system of certain interferences, the prior probabilities $\widetilde{P}^{(s)}$ at the initial moment of time will be equal to 1/S. As the prior probability $\widetilde{P}^{(s)}$ at the next step, its value from the previous step can be used. Under the assumption of a normal distribution of signals and noise, expression (7) can be represented as

$$\hat{P}_k^{(s)} = \left[\frac{\hat{P}_{k-1}^{(s)} \pi(J, Pr) exp[-0.5 g^{(s)}(z)]}{\sum_{s=1}^{S} \hat{P}_{k-1}^{(s)} \pi(J, Pr) exp[-0.5 g^{(s)}(z)]}\right]_k \tag{8}$$

where

$$g^{(s)}(z) = \sum_{i=1}^{I} g^{(s)(i)}(z^{(i)}), \ s = \overline{1,S}, \quad i = \overline{1,I}\,; \tag{9}$$

$$g^{(s)(i)}(z^{(i)}) = \left[\frac{z^{(i)} - \hat{x}^{(s)(i)}}{\sigma^{(i)}}\right]^2, \tag{10}$$

$z^{(i)}$, $\sigma^{(i)}$ — respectively the output signal of the i-th meter and its standard deviation; $\hat{x}^{(s)(i)}$ is the average value of the i-th phase coordinate in the s-th signal-interference situation.

The decision about the current signal-interference situation and, accordingly, the number of the structure of the information-measuring system of the moving object is made according to the criterion

$$\hat{s} = \underset{S}{argmax}(\hat{P}_k^{(s)}). \tag{11}$$

In the case when the a priori intensities $\tilde{\gamma}^{(ls)}, \tilde{\gamma}^{(sl)}$ of the change in signal-noise situations are known, expression (8) can be represented as

$$\hat{P}_k^{(s)} = \left[\frac{\tilde{P}^{(s)} \pi(J, Pr) exp[-0.5 g^{(s)}(z)]}{\sum_{s=1}^{S} \tilde{P}^{(s)} \pi(J, Pr) exp[-0.5 g^{(s)}(z)]}\right]_k, \tag{12}$$

where

$$\tilde{P}_k^{(s)} = \left[\hat{P}^{(s)} - \Delta t(\hat{P}^{(s)} \sum_{s-1 \neq l}^{S} \tilde{\gamma}^{(sl)} - \sum_{l=1 \neq s}^{S} \tilde{\gamma}^{(ls)} \hat{P}^{(l)}\right]_{k-1}, \Delta t = t_k - t_{r-1} \tag{13}$$

Thus the algorithm for recognizing signal-interference situations will consist of (12), (13), (6), (9)–(11). In this case, in the process of a moving object functioning, it is advisable to record the facts and duration of the effect of interference on the information-measuring system to correct the values $\tilde{\gamma}^{(ls)}, \tilde{\gamma}^{(sl)}$ used in (13).

Preliminary studies of the algorithm (12), (13), (6), (9)–(11) showed that when the a priori and actual intensities of the change in signal-noise situations coincide, the probabilities of making a correct decision about $\hat{s}$ are close to 1 and decrease as the mismatch grows.

Algorithms (8), (6), (9)–(11) and (12), (13), (6), (9)–(11) are quasi-optimal and depend on the accuracy of the used a priori data, which is typical for Bayesian approaches.

To improve the proposed algorithms, the opportunity of correcting the weight of a priori information in algorithm (12), (13), (6), (9)–(11) is considered in cases of its inconsistency with the actual data. The correction option consists in using a priori information about $\tilde{\gamma}^{(ls)}, \tilde{\gamma}^{(sl)}$ in expression (13) at the next step with a weight coefficient D depending on $\Delta\tau$

$$\tilde{P}_k^{(s)} = \left[\hat{P}^{(s)} - D\Delta t(\hat{P}^{(s)} \sum_{s-1 \neq l}^{S} \tilde{\gamma}^{(sl)} - \sum_{l=1 \neq s}^{S} \tilde{\gamma}^{(ls)} \hat{P}^{(l)})\right]_{k-1}, \tag{14}$$

where D=1/(d$\Delta\tau$+1); d — coefficient determined heuristically. The start of integration of time $\Delta\tau$, in addition to indicators, can be carried out when a predetermined threshold is reached for any phase coordinate of a moving object, selected when classifying signal-interference situations. To determine the correction parameters, it is necessary to perform statistical simulation of a specific moving object, its information-measuring system and signal-interference situations. Preliminary studies have shown that correction (14) can significantly improve the accuracy of algorithms (8), (6), (9)–(11) and (12), (13), (6), (9)–(11).

4. Conclusion

The paper proposes Bayesian algorithms for assessing the current state of a moving object and its information-measuring system, which make it possible to recognize various signal-interference situations from an a priori formed list. An approach to the correction of a priori data during operation is proposed. The algorithms are formulated in a very general form, which allows them to be applied to objects of various classes.

Acknowledgements

This work was supported by Russian Foundation for Basic Research (RFBR), project 20-08-00091-a, project 19-08-00487-a.

References

[1] Verba V S, Merkulov V I and Sadovsky P A 2019 Multipurpose tracking in multiposition radar systems *Journal of Communications Technology and Electronics* **64** (9) 973–980

[2] Boldinov V A, Bukhalev V A and Skrynnikov A A 2014 Recognition algorithm for the state of the queuing system based on theory of systems with random jump structure *Journal of Computer and Systems Sciences International* **53** (3) 327–337

[3] Boldinov V A, Bukhalev V A, Pryadkin S P and Skrynnikov A A 2016 Control of the probability distribution of a system state based on its structure indicator *Journal of Computer and Systems Sciences International* **55** (3) 333–340

[4] Bukhalev V A, Boldinov V A, Pryadkin S P and Skrynnikov A A 2015 Optimal control of linear stochastic system using smoothed estimate of phase coordinates *Automation and Remote Control* **76** (11) 1918–1930

[5] Pavlov V I 2017 Special features of forming the provoking probing signal in implementation of the semi-passive homing method for mobile objects *Russian Aeronautics* **60** (3) 374–381

IOP Conf. Series: Materials Science and Engineering 1027 (2021) 012022 doi:10.1088/1757-899X/1027/1/012022

Hydraulic losses optimization methods used in air conditioning system valve design for different closure angles

K S Napreenko and A V Lamtyugina

Moscow Aviation Institute (National Research University), 4 Volokolamskoe shosse, Moscow 125993, Russia

konaprko@yandex.ru

Abstract. This article describes methods for determining the hydraulic resistances of A/C system units. The advantages of computational hydrodynamics over the traditional method of calculation using the coefficient of hydraulic resistance are shown. A "step-by-step" example of calculating the hydraulic resistance of an A/C system unit in the ANSYS CFX software package is given (converting 3D geometry into a calculation model; selecting, configuring, and building a mesh model; configuring calculation parameters, including boundary conditions, turbulence models, and calculation stop criteria; calculating and processing results). Special attention is paid to the consideration of the SST turbulence model. The calculation results show the distribution fields of total and static pressure, velocity, and flow lines at discrete rotation angles. Graphically, the dependence of the drop in hydraulic resistance on the increase in the area of the flow section of the damper is shown. Hydraulic losses are also determined when the pressure increases. Conclusions are made about the application of the calculation results in creating a comprehensive model of the system

1. Introduction

Creating modern aircraft is a complicated, multi-iterative process. Currently, the development of aviation projects is carried out using the modeling method. After a set of models of individual systems and their properties is developed, it is possible to determine the appearance of the designed device [1–3].

Air conditioning system (A/C system) is one of the key units of the aircraft environmental control systems, which main task is to create conditions on Board the aircraft for normal human activity in flight. Any A/C system is a complex distributed multi-channel pneumatic system, the required operation of which is provided by the operation of the flow controller and the temperature of the supplied air. The A/C system consists of air extraction and cooling units, water extractors, filters, distribution equipment (adjustable valves, distributors, taps, etc.), connecting pipelines, etc. [4–6].

Part of the total energy is lost for this network irrevocably that goes to overcome the forces of hydraulic resistance that occurs when real gas moves through pipes and channels. The value equal to this irrevocable loss of total energy in this section is called the hydraulic resistance of the section [7].

For a comprehensive assessment of the system, it is important not only to know the main characteristics of the units but also to be able to calculate the hydraulic losses (the energy required to overcome the forces of hydraulic resistance) in each section.

One of the key units is the adjustable valve, which is designed to regulate the temperature of the air supplied to the cabin by regulating the flow of mixed air in the hotline of the pipeline. In addition to

IOP Conf. Series: Materials Science and Engineering 1027 (2021) 012022 doi:10.1088/1757-899X/1027/1/012022

the main characteristic (flow rate), an important parameter for selecting these aggregates is hydraulic resistance for different closure angles [4].

2. Method for calculating hydraulic losses

The method of estimating section hydraulic resistance by using the coefficient of hydraulic resistance (ζ) is widely known. This coefficient is defined as the ratio of the total pressure loss at the area (ΔP) to the value of the dynamic pressure at the area:

$$\zeta = \frac{\Delta P}{(\rho * \frac{w^2}{2})}$$

or

$$\Delta P = \zeta * (\rho * \frac{w^2}{2}) = \zeta * (\frac{\rho}{2}) * (\frac{Q}{F})^2 = \frac{\zeta}{\rho * 2 * (\frac{G}{F})^2}$$

where ρ — gas density in the investigated area, kg/m^3; w — velocity of the gas flow, m/s; Q — the volumetric flow rate of gas, m^3/s; F — a cross-section of the area, m^2, G — mass gas consumption, kg/s.

The values of the coefficient ζ for different types of hydraulic resistance sections are given in resource books or can be determined as a result of an experiment [7–9].

Along with using calculation methods based on criteria relations for hydrodynamics, it is advisable to develop calculation methods that use modern CFD (Computational fluid dynamics) codes. The completeness of the description of processes with such codes makes it possible to optimize the design and technological solutions, reasonably build physical modeling objects and experimental programs that are necessary for validating the calculated codes [10, 11].

In addition, also, when solving problems of hydraulics by using methods of computational hydrodynamics, it is possible to take into account the geometric features of the design, i.e. the calculation model takes into account the three-dimensional design of the calculation object [12].

Using modern design methods allows you to create a model of the system itself, and simulate its operation, thereby speeding up the process of creating a new aircraft, as well as optimize the cost of design, testing, and certification [13, 14].

3. Source data

The initial data for the calculation is the geometry of the valve and the parameters of the environment. The valve geometry is shown in Figure 1, the diameter of the flow area 80 mm. Environmental parameters: the working environment is air; temperature is 200 °C; pressure is one atm. Calculations of the open valve for the reference pressure of 5.5 atm were also made.

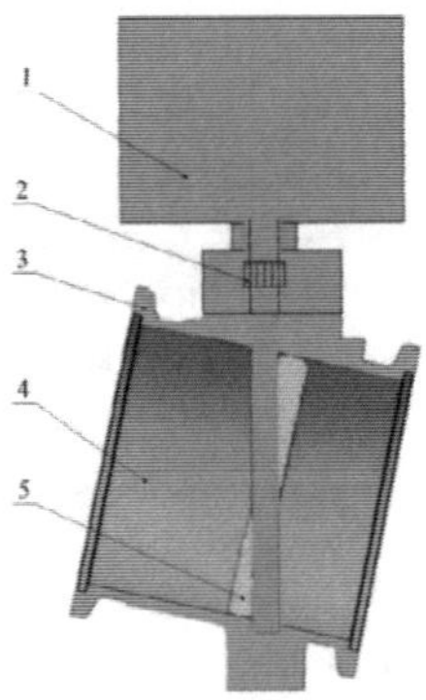

Figure 1. Three-dimensional model of the valve (pipeline diameter 80 mm). The numbers in figure 1 indicate 1 — the drive, 2 — the transmission mechanism, 3 — the connecting flanges, 4 — the airflow channel, 5 — the rotary part of the valve.

The calculation model is obtained from a three-dimensional model of the valve by "filling" the internal space of the valve elements with the environment and extending the channel to the sides of the air inflow and outflow on 10 diameters. Figures 2 and 3 show the design model of the valve, the angle of rotation of the flap is 30 degrees from the closed position.

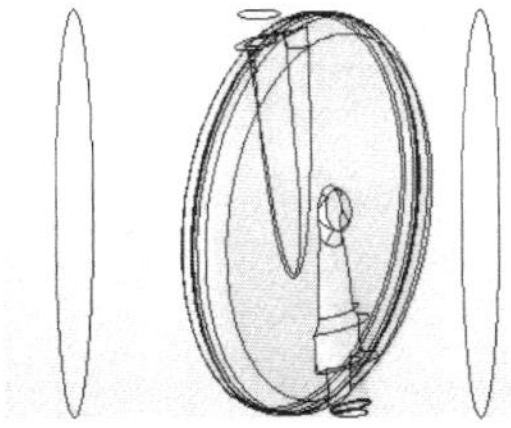

Figure 2. Geometry of the flow part of the valve.

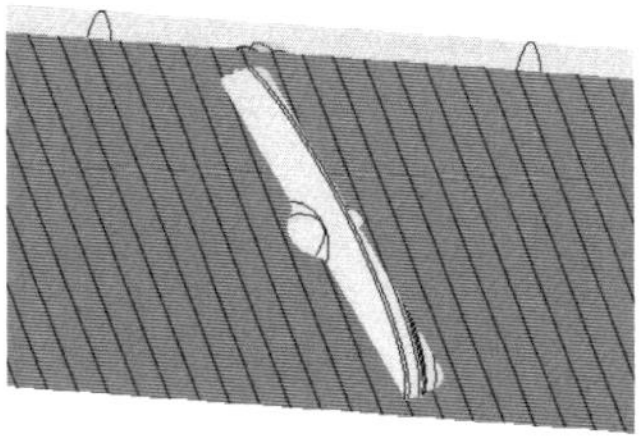

Figure 3. Geometry of the valve flow part (longitudinal section of the calculated model).

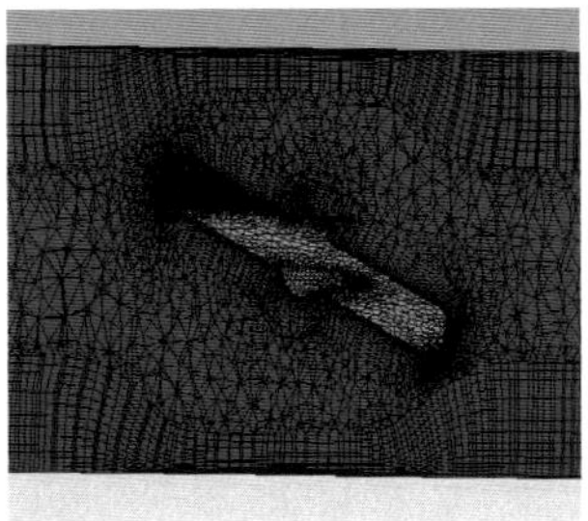

Figure 4. Computational mesh for the valve model.

The ANSYS CFX software package was used for the calculation. Ansys CFX is a software package that allows us to solve several problems in the field of computational fluid dynamics. The program uses the advanced algebraic multigrid conjugate solver. In the model settings, the solution of multiphase flows, combustion problems, rotary machines, chemically reacting mixtures, radiation, and a combination of conjugate calculations can be enabled [15–17].

4. Description of the selected turbulence model and environment parameters

The Shear Stress Transport (SST) turbulence model is used to model the flow inside the valve.

The SST turbulence model is a model for transferring tangent stresses. It is used when a good solution is required in the layer near the wall, which is important for calculating heat transfer.

The equations for the kinetic energy (k) and the turbulence frequency (ω) have the form:

$$\frac{\partial(\rho k)}{\partial t}+\frac{\partial}{\partial x_j}(\rho U_j k)=\frac{\partial}{\partial x_j}\left[\left(\mu+\frac{\mu_t}{\sigma_{k3}}\right)\frac{\partial k}{\partial x_j}\right]+P_k-\beta'\rho k\omega+P_{kb}$$

$$\frac{\partial(\rho\omega)}{\partial t}+\frac{\partial}{\partial x_j}(\rho U_j\omega)=$$

$$\frac{\partial}{\partial x_j}\left[\left(\mu+\frac{\mu_t}{\sigma_{\omega 2}}\right)\frac{\partial\omega}{\partial x_j}\right]+(1-F_1)2\rho\frac{1}{\sigma_{\omega 2}\omega}\frac{\partial k}{\partial x_j}\frac{\partial\omega}{\partial x_j}+\alpha\frac{\omega}{k}P_k-\beta\rho\omega^2+P_{\omega b},\ F_1=tanh(arg_1{}^4),$$

$$arg_1=min(max(\frac{2\sqrt{k}}{\beta'\omega y},\frac{500\nu}{y^2\omega}),\frac{4\rho k}{CD_{kw}\sigma_{\omega 2}y^2}),$$

$$CD_{kw}=max(2\rho\frac{1}{\sigma_{\omega 2}\omega}\frac{\partial k}{\partial x_j}\frac{\partial\omega}{\partial x_j},\quad 10^{-10})$$

where y — distance from the wall, m; α_2, β_2, σ_{k2}, $\sigma_{\omega 2}$ — constants, $\alpha_2 = 0{,}44$; $\beta_2 = 0{,}0828$; $\sigma_{k2} = 1$; $\sigma_{\omega 2} = 1/0{,}856$.

IOP Conf. Series: Materials Science and Engineering 1027 (2021) 012022 doi:10.1088/1757-899X/1027/1/012022

The SST model uses the "k-ω" model for the wall region (F_1=1) and the "k-ε" model for the external flow (F_1=0) [15, 18].

When modeling the near-wall flow, the following wall functions are used: standard, scalable, and automatic. For sufficient accuracy in describing the flow near the wall, a boundary layer (15–20 cells, with a growth coefficient of 1.2) was set, the first cell thickness of which corresponds to the value $y^+ = 10$ [19, 20].

5. Results of calculation

The hydraulic resistance of the valve at a temperature of 200°C is summarized in the table 1 and graph (Figure 5).

Table 1.

Flap rotation angle, degrees	ΔP, Pa
25	420000
30	250000
35	165000
40	100500
45	59811
50	33306
55	18749
60	11033
65	6692
70	4286
75	2857
80	2100
85	1633
90	1392

Figures 5, 6 show the results of calculations for angles of flap rotation 90° and figures 7, 8 for 75°. The drawings also contain information about the distribution patterns of excess static and total pressure, velocity, velocity vectors, and flow lines.

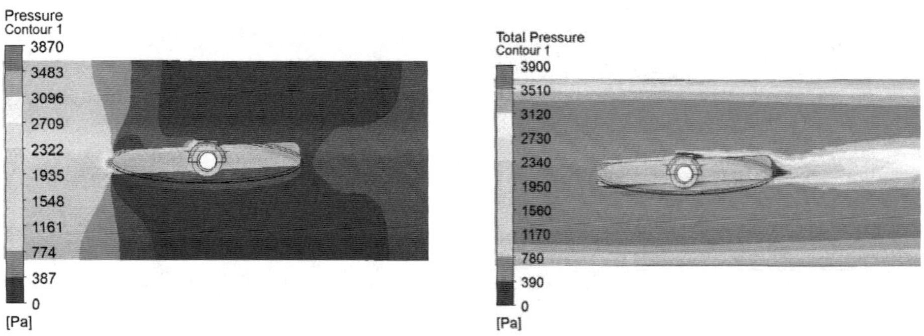

(a) (b)

Figure 5. Pressure distribution, angle of flap rotation 90° (a) — excessive static pressure; (b) — excessive total pressure).

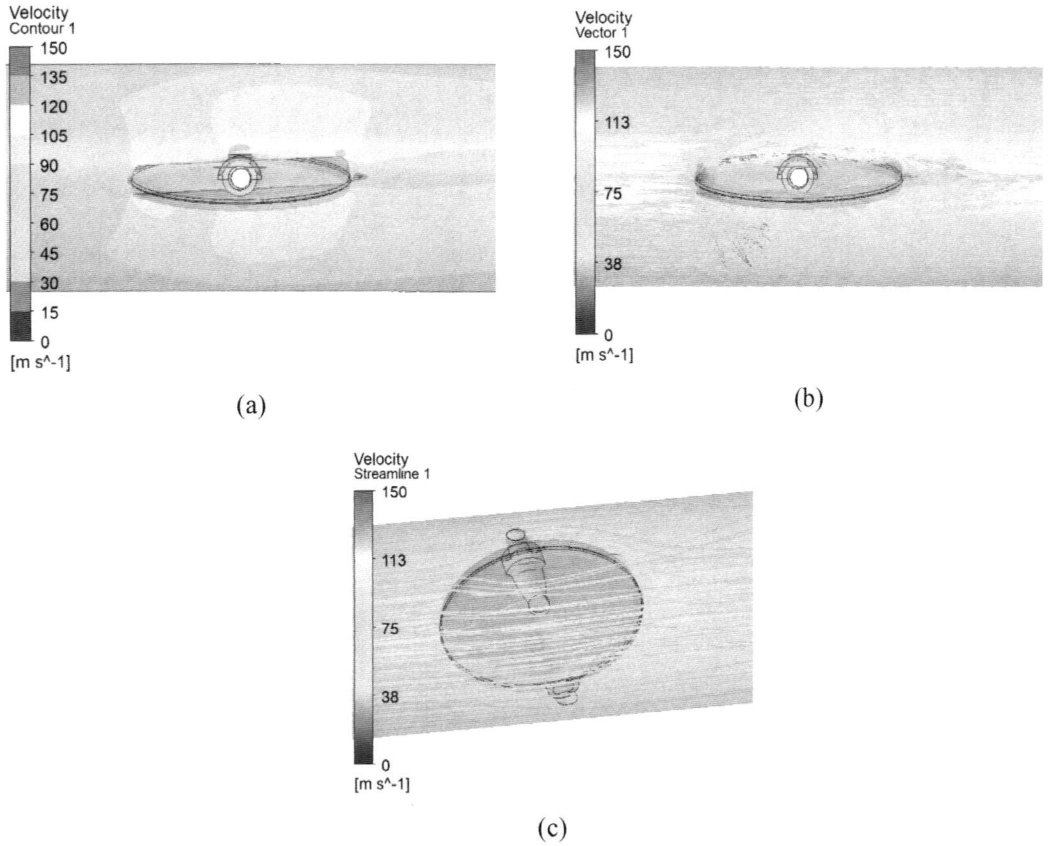

Figure 6. Distribution of velocity, angle of flap rotation 90°. (a) — velocity, (b) — velocity vector, (c) — flow lines.

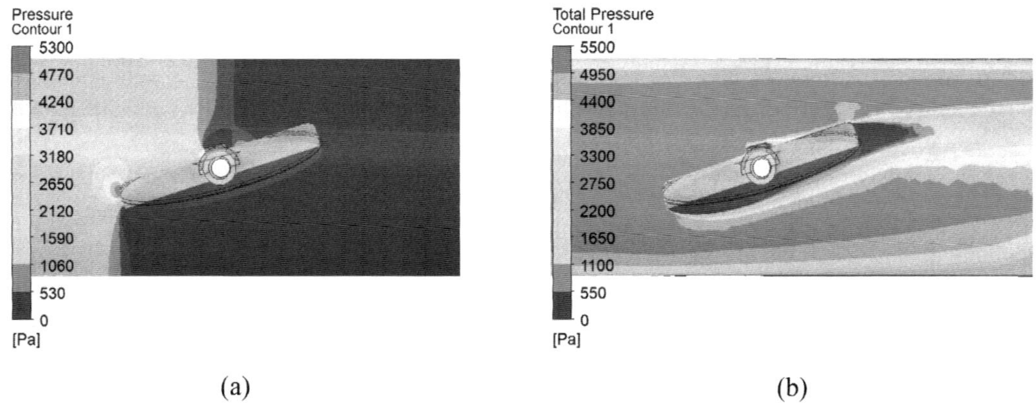

Figure 7. Pressure distribution, angle of flap rotation 75°. (a) — excessive static pressure, (b) — excessive total pressure.

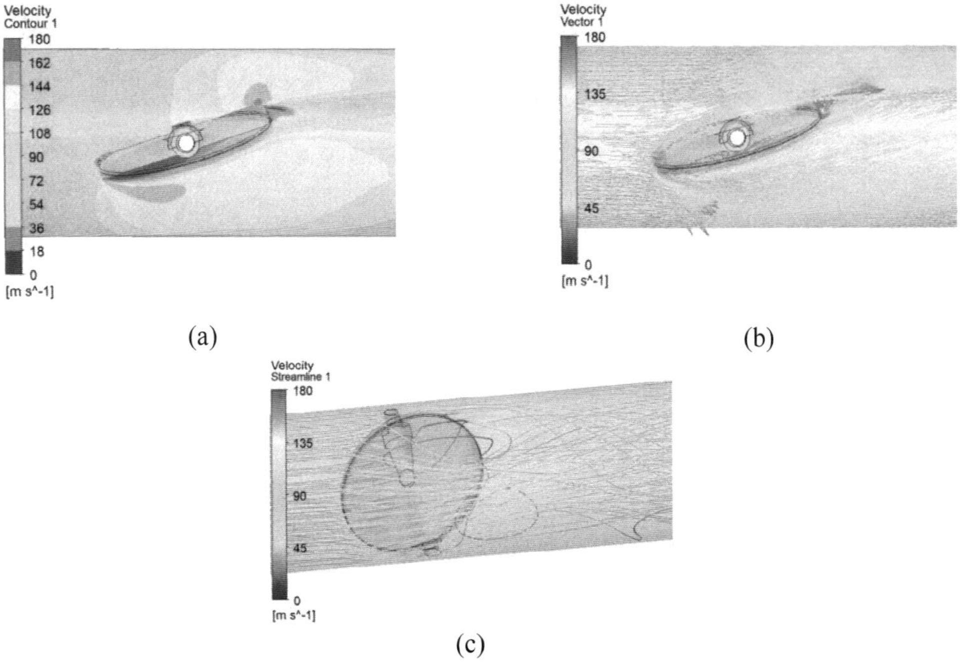

Figure 8. Distribution of velocity, angle of flap rotation 75°. (a) — velocity, (b) — velocity vector, (c) — flow lines.

6. Conclusion

The method for calculating the hydraulic resistance of the valve described in this paper can also be applied to other air conditioning units. After performing all calculations and validating them, the found values of hydraulic resistance should be included in the general model of the A/C system, thus, creating a complete and reliable picture of the pressure distribution in the system.

References

[1] Smagin D I, Satin A A, Saveliev R S, Starostin K I, Suzdaltseva D S and Serebryansky S A 2018 The concept of the creating an integrated mathematical model of the aircraft including the main onboard systems *J. Quality and life* **4** (20) 397–403

[2] *Aircraft engineering* ed M A Pogosyan. 5th ed., revised and enlarged edition (Moscow Innovative mechanical engineering) p 864

[3] Pogosyan M A, Savelievskikh E P, Shagiliev R M and Kozelkov A S 2013 *Problems of atomic science and technology Ser. Mathematical modeling of physical processes* **2**

[4] Voronin G I 1973 *Air conditioning systems on the aircraft*: *textbook* (Moscow: Mechanical Engineering)

[5] Kravchenko A G, Shilova A K, Tamba-Tamba V P and Ozersky AI 2017 Review of the main units of the air conditioning system of the aircraft *J. Technics. Technology. Engineering* **3** 24–7

[6] Akopov M G, Bekasov V I and Dolgushev V G and others 2005 *Aircraft equipment systems*: *Textbook for C40 students of the higher technical educational institutions* ed A M Matveenko and V I Bekasova 3rd ed. (Moscow: Mechanical Engineering)

[7] Idelchik I E 1992 *Handbook of the hydraulic resistance* ed M O Steinberg. 3rd ed., revised and enlarged edition. (Moscow: Mechanical Engineering) p 672

[8] Altushul A D, Kalitsun V I and Mayranovsky F G etc. 1976 *Examples of the calculations for hydraulics*: *Textbook* (Moscow: Stroyizdat) p 256

[9] Kireev V, Nizamova A and Urmancheev S The hydraulic resistance of thermoviscous liquid flow in a plane channel with a variable cross-section *J. of Ph.: Conf. Series* **1158** 3

[10] Borzenko E I, Ryltseva K E and Schrager G R 2019 *Bulliten of Tomsk State University Mathematics* Numerical study of the characteristics of the flow of a non-Newtonian fluid in demand with a sudden contraction **58** 56–70

[11] Afremov D A, Afonin S Yu, Zakharov A G, Nedayvozov A V and Smirnov V P 2015 Verification of three-dimensional CFD codes for calculating the forced flow in fuil assemblies of fast reactors *NIKIET Annual Report – 2015* Ed E O Adamova. (Moscow) p 154

[12] Anikeev A A, Molchanov A M and Yanyshev D S 2010 *Fundamentals of computational heat transfer and hydrodynamics. Textbook* (Moscow)

[13] Federal Aviation Regulations. Part 25 Airworthiness Standards: Transport Category Airplanes

[14] Federal Aviation Regulations. Part 29 Airworthiness Standards: Transport Category Rotorcraft

[15] Ansys help, ansyshelp.ansys.com

[16] Methods for the Design of Modern On-Board Systems of Advanced Aircraft. 2019 IEEE *10th Int. Conf. on Mech. and Aerospace Eng. (ICMAE)* 2019 DOI: 10.1109/ICMAE.2019.8880986

[17] Molchanov A M 2019 *Numerical method for solving the Navier-Stokes equations* OSF Preprints, Web

[18] Badernikov A V, Piralishvily S A and Guryanov A I 2018 Results of Numerical Modeling of Combustion Processes in a Vortex Chamber *MATEC Web of Conf.* **209** 00023

[19] Benner P, Gugercin S and Willcox K 2015 *A survey of projection-based model reduction methods for parametric dynamical systems* SIAM Rev. 57

[20] Vreman A W, Sandham N D and Luo K H 1996 Compressible mixing layer growth rate and turbulence characteristics *J. of Fluid Mech.* **320** 235–58

IOP Conf. Series: Materials Science and Engineering 1027 (2021) 012023 doi:10.1088/1757-899X/1027/1/012023

Control synthesis for tasks with a given output function decentralized SISO-systems

E Nikulchev

MIREA – Russian Technological University, Moscow 119571, Russia

nikulchev@mail.ru

Abstract. For the control synthesis of multiply connected objects with many inputs and many outputs (MIMO) by importance of reducing computational costs by reducing accuracy, the method of decentralized control is used by dividing the original system into systems with one input and one output (SISO). The paper proposes an approach to the control synthesis of a decentralized SISO system, which is a part of a multi-connected MIMO object in conditions of impossibility of identifying model parameters. The paper considers the problem of control synthesis, which provides a given desired form of the output process.

1. Introduction

Large technical control systems in industry are described based on models with many inputs and outputs (MIMO), sometimes they have very large dimensions. Although there are modeling techniques and approaches to control synthesis, the computational costs of optimization problems are enormous. An alternative approach is a decentralized management strategy, splitting in some way into systems with one input and one output (SISO) [1]. The loss of accuracy and other efficiency criteria in decentralized control versus full multi-parameter control is compensated by the simplicity of the equipment, ease of development and use. The basic level of regulatory control in MIMO installations mainly consists of SISO decentralized PID controllers [2] due to their simple structure, fewer tunable parameters and loop fault tolerance, since it is easier to identify and eliminate sensor or actuator failure in a decentralized structure, although it includes a significant number of specialized tasks [3].

An important feature of splitting into SISO systems is the complexity of identification, since it is difficult to send test signals directly to a selected object, for example, if the object is part of a large mechanism or automatic line and is not directly controlled by the input parameters of the entire MIMO system. However, in practice, the projector of the control system can formulate the requirements for a specific decentralized SISO-system in the form of the desired schedule, which the system should issue in the specified modes. In practice, there have been cases when, due to the design features and technical requirements of the technological process, it is impossible to send a test signal for object identification, and only smooth signals are available (for example, in heating systems; in systems with sensitive to sudden changes in parameters), while the desired type of the output function is set. In these cases, it is needed to find the kind of input function that provides the required output. Control kind found can be provided by Iterative Learning Control [4], repetitive control [5] or fuzzy controller technologies.

It should be noted that the approach of searching for transformations that leave the required characteristics invariants is widely used in radio electronics [6]. Solving the problems of searching for mapping admitted by differential equations allows to obtain methods for modeling objects with nonlinear elements [7], generating robust chaos [8] and systems with hidden attractors [9], and others. In this paper, it is proposed to replace the identification of a SISO object of a multiply connected MIMO system with a search for the type of control action of this SISO object, which provides the required specified type of output signal. It is proposed to look for a solution in the form of graph transformation.

2. Method

Let there be a SISO LTI system described by the differential equation:

$$\begin{aligned} \dot{x} &= Ax(t) + Bu(t), \\ y(t) &= Cx(t), \end{aligned} \tag{1}$$

where $x(t)$ — n-dimensional vector of systems states from a variety X; $u(t)$ — control from $\mathbb{R}$; $y(t)$ — output measured process from $\mathbb{R}$; X — variety, usually identified with $\mathbb{R}^n$, or an invariant torus; $\dot{x} = (dx)/(dt) \in T_x X$; $T_x X$ — tangent space to X in x, determined by control; $A \in \mathbb{R}^{n\times n}$; $B \in \mathbb{R}^{n\times r}$; $C \in \mathbb{R}^{m\times n}$.

Let for fixed $u^1(t)$ and $u^2(t)$ corresponding processes are observed $y^1(t)$ and $y^2(t)$. Then if the symmetry group G is allowed by system (1), then the transformation g:

$$g : y^1(t) \to y^2(t),$$

is at the same time a transformation between control processes:

$$g : u^1(t) \to u^2(t).$$

As an illustration of the proof, the communication scheme is presented:

$$\begin{array}{ccc} u^1 & \xrightarrow{\xi} & y^1 \\ {\scriptstyle g}\downarrow & & \downarrow{\scriptstyle g} \\ u^2 & \xrightarrow{\xi} & y^2 \end{array}$$

Here ξ is transformation that transfers the control process to the observable output process. It is given by the system of equations 2 with the same initial conditions.

In other words, in the system under control $u^1(t)$, output processes $y^1(t)$ are observed in accordance with the system of equations (1). By controlling $u^1(t)$, output processes $y^1(t)$ are observed respectively. Since the solutions of system (1) form a Lie group, that is, a group of transformations with a commutation operation, then there is a transformation that transforms one solution into another. By the Grobman-Hartman theorem, such a transformation is linear.

The solution to stationary system (1) is defined as following

$$y(t) = \int_{t_0}^{t} C\Phi(t-\tau)Bu(\tau)d\tau,$$

where $\Phi(t-\tau) = \exp\{A(t-\tau)\}$.

For the output process

$$y^1(t) = \int_{t_0}^{t} C\Phi(t-\tau)Bu^1(\tau)d\tau,$$

by condition

$$y^2(t) = g(y^1(t)),$$

from

$$y^2(t) = g\left(\int_{t_0}^{t} C\Phi(t-\tau)Bu^1(\tau)d\tau\right),$$

Considering the G group properties, following is found

$$y^2(t) = \int_{t_0}^{t} C\Phi(t-\tau)Bg\left(u^1(\tau)\right)d\tau, \tag{2}$$

on the other hand, the solution for the output process y^2:

$$y^2(t) = \int_{t_0}^{t} C\Phi(t-\tau)Bu^2(\tau)d\tau, \tag{3}$$

Thus, from the uniqueness of the process y^2 defined by equations (2) and (3), it follows that

$$u^2(t) = g\left(u^1(t)\right).$$

Thus, it is shown that to find the desired function u^2 that provides a given form of output y^2, it is sufficient to find the transformation g and apply it to the observed form u^1. The conversion g is looked for as the conversion of the desired function to the given one.

3. Numerical experiment

Figure 1 shows two output processes: system characteristic (blue), desired process (orange)

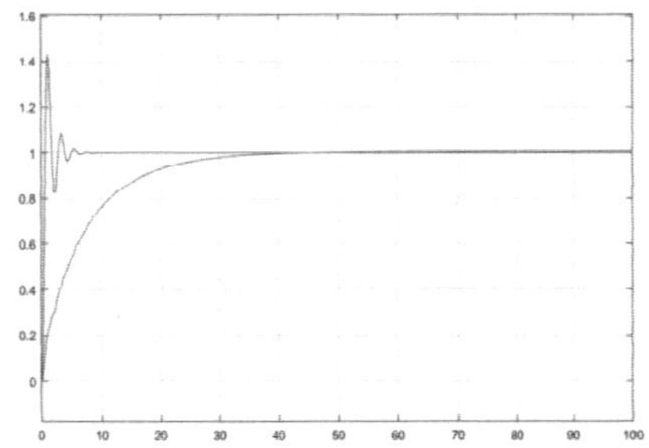

Figure 1. Observed and Desired System Output Functions.

An approximation of the transformation g in the form of a polynomial was found [10]. When executing the program control scheme shown in Figure 2, a view of the output process corresponding to the given one is obtained. The discrepancy is due to the error of the numerical methods used.

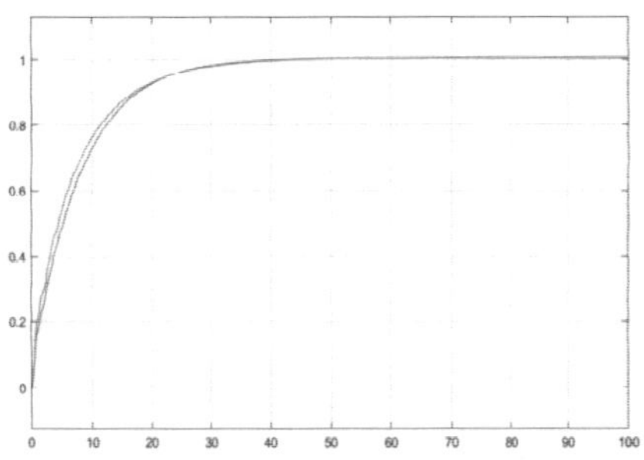

Figure 2. System output with found control (blue — given output, orange — observed system output).

4. Conclusions

The paper proposes an approach to the synthesis of control of a decentralized SISO system, which is a part of a multi-connected MIMO object in conditions of impossibility of identifying the parameters of an LTI object. The problem of such a control synthesis, which provides a given desired form of the output process, is considered. It is shown theoretically and numerically, that to find the desired input function that provides a given type of output, it is enough to find a transformation and apply it to the observed type of the input process.

References

[1] Euzébio T A, Yamashita A S, Pinto T V and Barros P R 2020 *Journal of Process Control* **94** 75-96
[2] Boyd S, Hast M and Astrom K J 2016 *Internat. J. Robust Nonlinear Control*, **26** 1718-1731
[3] Yin H, Lin S, Ye F and Zhang W 2020. *Asian Journal of Control* **22** 831-840
[4] Li X, and Ruan X 2019 *Advances in Difference Equations* **2019** 59.
[5] Kurniawan E, Cao Z W, Mitrevska M and Man Z H 2016 *International Journal of Automation and Computing* **13** 615-623.
[6] Smirnov A V 2020 *Russian Technological Journal* **8**(1) 43-58 doi: 10.32362/2500-316X-2020-8-2-43-58
[7] Nikulchev E and Kondratov A 2015 *Advanced Studies in Theoretical Physics* **9**(16) 795-800
[8] Nikulchev E 2014 Robust chaos generation on the basis of symmetry violations in attractors. *2014 2nd 2014 2nd International Conference on Emission Electronics (ICEE)*, pp. 1-3 doi: 10.1109/Emission.2014.6893972
[9] Dudkowski D, Jafari S, Kapitaniak T, Kuznetsov N V, Leonov G A and Prasad A 2016 *Physics Reports* **637** 1-50.
[10] Nikulchev E 2014 *Advanced Studies in Theoretical Physics* **8**(21) 939-944

IOP Conf. Series: Materials Science and Engineering 1027 (2021) 012024 doi:10.1088/1757-899X/1027/1/012024

Surface microgeometry monitoring of large-sized aircraft elements

V I Pronyakin[1], V L Skrypka[1] and B T Abykanova[2]

[1] Bauman Moscow State Technical University, 5 Baumanskaya 2-ya, Moscow 105005, Russia
[2] Kh.Dosmukhamedov Atyrau University, 1 Student Avenue, Atyrau E01Y6PO, Kazakhstan

vip-u@bmstu.ru

Abstract. The article is devoted to the assessment of the surface microgeometry of structural elements of an aircraft with a significant surface area. A statistical method for assessing the surface quality using correlation methods for analyzing random processes is presented. An example of processing results is given.

1. Introduction

One of the tasks solved during the design and operation of an aircraft is to ensure its aerodynamic characteristics, flight safety and fuel economy [1]. The microgeometry of the aircraft structural elements surfaces has a significant influence on these characteristics. The aircraft surfaces structure is largely determined by the widespread use of polymer and composite materials in aircraft construction. The surface microgeometry also depends on the surface protection against the external environment (multilayer, erosion-resistant, fabric-film, paint, varnish and other coatings). At the same time, the current methods for assessing the characteristics of surfaces are based on the analysis of small areas, which results are extrapolated to the entire rest of the surface [2]. Such an assumption in the surface survey fails to identify defects located outside the area being evaluated. The overwhelming majority of primary measuring transducers (mechanical, optical, eddy current, magnetic, etc.), used to control the deviation of the geometry of surfaces, have largely exhausted their capabilities to improve the accuracy [3]. Therefore, it is promising to improve algorithms for computer processing of measurement data.

2. Statistical method for assessing surface quality

Since the microgeometry of surfaces is a realization of a random process, statistical methods of analysis of random processes can be used for processing. The feasibility of implementing this approach has been convincingly proven in radio engineering [4, 5]. Reasoning by analogy, it can be assumed that, for example, correlation processing of measuring signals [6] can significantly increase the metrological characteristics of monitoring systems.

The aircraft outer surfaces defects may be classified as follows:

- roughness, that is, a collection of surface irregularities with relatively small steps;

- dirt, that is, particles of solid substances (dust, sand, etc.) adhering to the surface, the distribution of which is irregular;
- microcracks formed during the operation due to aging and erosion, which are depressions with relatively large steps;
- waviness, that is, macroroughness on a surface, characterized by smooth transitions from protrusions to depressions;
- protrusion or sinking of the heads of rivets, bolts, screws;
- protrusion of welded seams, dents of resistance welding, skin tightening along riveted seams;
- ledges and gaps along sashes, hatches, etc.

Roughness and waviness have the greatest influence on the quality of the aircraft surface [7, 8].

The control of the microgeometry of the aircraft surface, based on the use of correlation analises, is advisable to be carried out using paired optical means that can provide the required speed. In the case when a defect appears on the surface, the statistical characteristics in this zone will be different than that of a normal quality surface, which will cause a change in the correlation function of the signals of the two sensors. Therefore, by successively comparing the correlation functions of the measurement signals corresponding to the control of neighboring areas of the surface, it is possible to estimate the degree of its homogeneity and manifestation of defects [8].

Let the generated measuring signal $S(x)$ represent the sum of the signals $y(x,\lambda(x)$, depending on one parameter $\lambda(x)$, characterizing the surface structure (for example, the height of irregularities), and stationary white noise $n(x)$:

$$S(x) = y(x, \lambda(x)) + n(x),\ 0 \le x \le L. \tag{1}$$

In this case, the assessment of the change in the structure of the signal $y(x,\lambda(x)$, caused by the presence of a surface defect, is largely determined by the accuracy of its estimation. For the criterion of the accuracy of the estimation of random signals, it is advisable to use the maximum of the posterior probability of the value of the parameter $\lambda(x)$ [8].

In this case, if $\lambda(x)$ over the observation interval $(0 \le x \le L)$ is a simple Markov process, then for surface elements x and $x+\Delta x$ located close to each other (with Δx much smaller than the size of the surface defect), the posterior probability density f_S for the values of the parameter $\lambda(x)$ on these two adjacent intervals is [4]:

$$\begin{aligned}\frac{\partial f_S(x,\lambda)}{\partial x} &= \frac{1}{2}\cdot\frac{\partial^2}{\partial\lambda^2}\cdot[Q_2(\lambda)f(x,\lambda)] - \frac{\partial}{\partial\lambda}\cdot[Q_1(\lambda)f(x,\lambda)] + \\ &+[F(x,\lambda)-\overline{F}(x,\lambda)]\cdot f(x,\lambda),\end{aligned} \tag{2}$$

where: $F(x,\lambda) \approx \frac{2}{N_0}\cdot S(x)\cdot y(x,\lambda)$; N_0 — spectral density of white noise; $\overline{F}(x,\lambda)$ — average value on sites; Q_1, Q_2 — coefficients describing the "behavior" of the parameter $\lambda(x)$;

$Q_1(l) = \lim\limits_{l\to 0}\frac{\overline{(\lambda_l-\lambda)^2}}{l}$; $Q_2(l) = \lim\limits_{l\to 0}\frac{\overline{(\lambda_l-\lambda)^2}}{l}$.

Thus, the accuracy is determined by the value $\lambda_m(x)$, corresponding to the maximum of the posterior probability distribution (2), and by the variance of this distribution. The practical implementation of algorithm (2) can be very difficult due to the inaccuracy of the real values of the input parameters. At the same time, if we assume that the posterior probability density is Gaussian, that is,

IOP Conf. Series: Materials Science and Engineering 1027 (2021) 012024 doi:10.1088/1757-899X/1027/1/012024

$$f_S(x,\lambda) = \frac{1}{\sqrt{2\pi\sigma_\lambda^2(x)}} \exp\left\{-\frac{[\lambda(x)-\overline{\lambda}(x)]^2}{2\pi\sigma_\lambda^2(x)}\right\},$$

then it is sufficient to estimate only the mean value $\lambda(x)$ and $\sigma_\lambda^2(x)$, which characterize the confidence interval of the posterior distribution. In this case, expression (2) can be reduced to the form [4, 8]

$$\frac{d\overline{\lambda(x)}}{dx} = Q_1(\overline{\lambda}) + \sigma_\lambda^2 \cdot F'(x,\overline{\lambda}).$$

$$\frac{d\sigma_\lambda^2}{dx} = 2\sigma_\lambda^2 \cdot Q_1(\overline{\lambda}) + Q_2(\overline{\lambda}) + \sigma_\lambda^4 \cdot F''(x,\overline{\lambda}).$$

The joint solution of the last equations gives an optimal estimate of $\lambda(x)$. The assumption of a normal distribution $f_S(x,\lambda)$ will be justified at large signal-to-noise ratios for which the condition

$$\sigma_\lambda(x) \square \Delta\lambda,\ \Delta\lambda \square \frac{y(x,\lambda)}{\dfrac{\partial y(x,\lambda)}{\partial\lambda}}.$$

When assessing the quality of the surface microgeometry, it is proposed to use one generalized parameter — the correlation function between adjacent areas of the surface or the controlled area and the sample.

The scheme for the formation of such an assessment of the quality of the wing surface is shown in Figure 1. The structure of the aircraft surface is represented by a random spatial process S (x), the parameters of which are measured in zones x_1 and x_2 by sensors A_1 and A_2 at a fixed distance ("step"), which have the corresponding output signals $S_1(x)$ and $S_2(x)$.

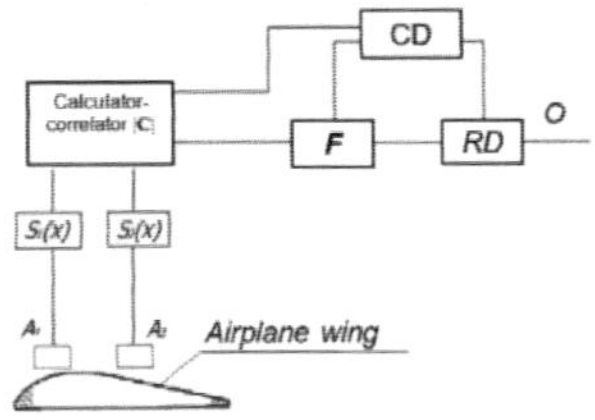

Figure 1. Arrangement of primary measuring transducers for correlation signal processing. UU — control device; F — high-pass filter; RD — regulating device; OD — output data; | C | — correlator; A — primary measuring transducer (sensor).

When step-by-step movement of the primary measuring transducers for a time sufficient for signal processing using a computer, auto- or cross-correlation functions are determined, which are the "carrier" of detailed information about the structural properties of the surface, and about the statistical characteristics of the signals $S_1(x)$ and $S_2(x)$ and interference.

The results of the parameter estimation are used for detection and recognition of measuring signals carrying information about the actual surface structure. This method of processing the received measurement signals should allow to detect the presence or absence of a defect.

If we assume that the measuring signal for the surface, with the required degree of quality, is known, then the implementation of the defect detection rule leads to a scheme for detecting the known signal $S_0(l)$ against the background of noise $n(l)$, where l — is the distance between the sensors [8].

This method of surface quality control uses an integral assessment of the characteristics of microroughnesses from statistical surface profilograms. For this, non-contact and contact methods can be used, for example, the method of interference of the incident and reflected light flux, raster and reflectometric methods, the method of diagnosing with a light beam, etc.

To increase the reliability of control, it is advisable to proceed from one-dimensional to two-dimensional correlation functions [8].

To determine the two-dimensional correlation function in the field of interaction of the primary measuring transducer with the surface, a strip with a width of N pixels is allocated and a digital reference pattern of size $N \times N$ is attributed to the center of this strip. Then the reference pattern and the highlighted band are sent to the correlator. As a reference pattern, both the previous value of the $S(x)$ signal and the signal from the reference surface can be used. When the reference pattern is virtually moved along the selected strip with a certain step and each alignment with the current fragment of the image, the correlator calculates the number of matching pixels between them. This sum is considered as the autocorrelation coefficient. To obtain the normalized value of the autocorrelation coefficient K, the resulting sum is divided by the area of the control sample. The found K values are stored in the computer memory. After completing the calculation of the autocorrelation coefficients of the first band, the next band is controlled in the same way as the previous one, but shifted by a few pixels. In this band, the previous reference pattern is used to calculate the autocorrelation coefficient and the steps are repeated, etc. Thus, after processing the entire surface, a matrix of autocorrelation coefficients or a two-dimensional discrete autocorrelation function is formed. It is used for quantitative assessment, on the basis of which it is possible to reliably recognize the deviation of the studied surface from the norm with a given probability. For the autocorrelation matrix, the average value of the autocorrelation coefficient K_{mean} is determined. Then, the found K_{mean} value is subtracted from each element of the autocorrelation matrix and the sum of all differences in absolute value is calculated. This sum is then divided by the size of the autocorrelation matrix **D**:

$$U = \frac{\sum_{i=1}^{D} | K - K_{mean} |}{\mathbf{D}}.$$

The thus obtained estimate of U can be considered as the average amplitude of the variable component of the two-dimensional autocorrelation function. The calculated value of U with a given probability P is used to determine the confidence interval in which this random variable falls. Then, using the found boundaries of the confidence interval, the parameters of the controlled surface are estimated [8].

3. Surface scan results.

Thus, the use of algorithms for correlation processing of the measurements of the aircraft surface structure increases the reliability of control without changing the design of the used measuring instruments. For example, the PIK-30M interference computer profilometer [6] makes it possible to measure fragments (up to 150 mm) of large parts with a surface reflection coefficient of more than 4% and a radius of curvature from 1.5 to 8 meters. Figure 2 shows the characteristic changes in the values of the correlation functions of the measurement signals obtained by scanning samples with a “rough” roughness (sample 1) and a “smoother” surface (sample 2).

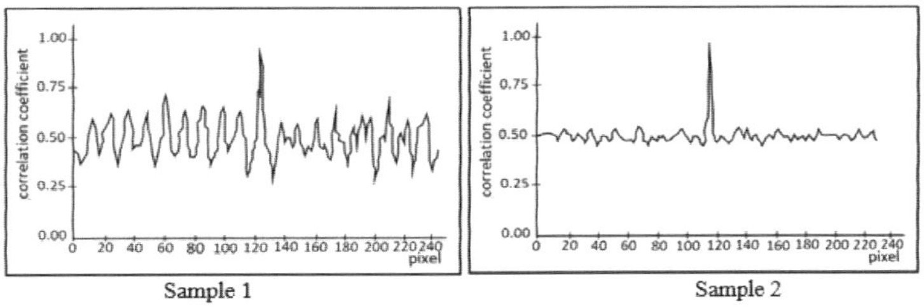

Figure 2. Graphs of changes in the normalized correlation signals of the investigated surfaces.

4. Conclusion

Thus, the use of correlation algorithms for t processing the measurement data increases the reliability of surface monitoring using the currently available technical means. In addition, using ultrasonic or magnetic sensors, it is possible to detect not only surface, but also subsurface defects, and find the relationship between them. The presented approach to surface quality monitoring is applicable for implementation both in manufacturing conditions and in aircraft operating conditions.

References

[1] *Aerodynamics, stability and controllability of supersonic aircraft* 1998 ed G S Bjushgens (Moscow: Nauka) p 816
[2] Whitehouse D J 2003 *Handbook of surface and nanometrology* (IOP Publishing Ltd)
[3] Badiru A B, Valensia V V and Liu D 2017 *Additive Manufacturing Handbook: Product Development for the Defense Industry* (CRC Press, Taylor & Francis Group) p 924
[4] Tichonov V I 2012 *Statistical radiotechnics* (Moscow: EE Media) p 678
[5] Khokhlov V K 2005 *Detection, recognition, and direction finding of objects in a nearby location* (Moscow: Bauman Moscow state technical University publishing house) p 334
[6] Makeev A V 2014 Analysis of modern methods for studying the surface roughness of parts *The Siberian State Geodetic Academy Bulletin* **4** (28) 80–7
[7] Wang S H, Quan C, Tay C J and Shang H M 2000 Surface roughness measurement in the submicrometer range using laser scattering *Optical engineering* **39** 6
[8] Abramov A D 2008 Determination of the roughness of polished surfaces based on the analysis of their autocorrelation functions *Proceedings of the Samara scientific center of the Russian Academy of Sciences* **10** (3) 887–994

IOP Conf. Series: Materials Science and Engineering 1027 (2021) 012025 doi:10.1088/1757-899X/1027/1/012025

Criterion and principles of easily-modified program

A Prutzkow[1,2]

[1] Department of Computational and Applied Mathematics, Ryazan State Radio Engineering University named after V. F. Utkin, 59/1 Gagarina str., Ryazan 390005, Russian Federation
[2] Department of Mathematics, Physics, and Medical Computer Sciences, Ryazan State Medical University named after I. P. Pavlov, 9 Vysokovoltnaya str., Ryazan 390026, Russian Federation

mail@prutzkow.com

Abstract. Ease of modifying is one of the characteristics of the program. The existing approaches do not clarify this concept. These approaches are clean code and maintainability. We formulate simple and explicit criterion and principles of easily-modified programs. The criterion is an E.W. Dijkstra's criterion. The principles are (1) the names of the elements should reflect their purpose in the program; (2) use indentation for nested statements and blank lines; (3) one program element must have one assignment and one assignment must correspond to only one element; (4) optimize dependencies between program elements. We introduce a metric for easily-modified programs as well. It's a number of commands in methods and static blocks in a class.

1. What is a Good Program?

Usually there are many programs to solve a problem. How to choose a good program? To say that one program is better than another, you need to choose a comparison criterion. Programs can be compared according to various characteristics: performance, the amount of required computer resources, the length of its program text, usability, etc.

Changes to the program do not end after the developer transfers it to the client. The less time it takes to make changes to the program, the cheaper it is to maintain. Therefore, an important characteristic of the program is the ease of modifying it.

There are two approaches to clearify this concept. These are clean code and maintainability.

2. Approaches Clearifying Ease of Modifying Program

2.1. Clean Code

Original source of clean code is the R. Martin's book [1]. In the «What is Clean Code?» section he collected some citations of famous programmers about clean code. They are following:

(1) «Clean code is simple and direct. Clean code reads like well-written prose. Clean code never obscures the designer's intent but rather is full of crisp abstractions and straightforward lines of control.»

(2) «Clean code can be read, and enhanced by a developer other than its original author. It has unit and acceptance tests. It has meaningful names. It provides one way rather than many ways for doing one thing. It has minimal dependencies, which are explicitly defined, and provides a clear and minimal API. Code should be literate since depending on the language, not all necessary information can be expressed clearly in code alone.»

(3) «I could list all of the qualities that I notice in clean code, but there is one overarching quality that leads to all of them. Clean code always looks like it was written by someone who cares. There is nothing obvious that you can do to make it better. All of those things were thought about by the code's author, and if you try to imagine improvements, you're led back to where you are, sitting in appreciation of the code someone left for you – code left by someone who cares deeply about the craft.»

(4) «You know you are working on clean code when each routine you read turns out to be pretty much what you expected. You can call it beautiful code when the code also makes it look like the language was made for the problem.»

Clean code determines the quality of the software [2]:

(5) «There are several reasons why you should care about code quality.

First, the quality of your code will determine the quality of the software. If you think you can have great software with messy code, you're dreaming. Second, clean code saves time. You will spend more time reading your code more than you spend writing it. The other people in your team will also spend a good part of their time reading your code. And you will spend time reading other people's code too. Clean code is easy to read, so clean code saves time. Third, your code is a reflection of yourself. If you don't value quality in your code, you won't get hired. Messy code means you don't hold yourself, or your work, to a very high standard. Clients and employers demand quality.

Clean code is maintainable. It is easy to read and easy to understand. That means it's easy to use and easy to change. Clean code is straightforward.»

Citation is from [3]:

(6) «There is no sole or strict definition of clean code. Moreover, there is probably no way of formally measuring clean code, so you cannot run a tool on a repository that could tell you how good, bad, or maintainable or not that code is. Sure, you can run tools such as checkers, linters, static analyzers, and so on. And those tools are of much help. They are necessary, but not sufficient. Clean code is not something a machine or script could tell (so far), but rather something that us, as professionals, can decide.

For decades of using the terms programming languages, we thought that they were languages to communicate our ideas to the machine, so it can run our programs. We were wrong. That's not the truth, but part of the truth. The real language behind programming languages is to communicate our ideas to other developers.

Here is where the true nature of clean code lies. It depends on other engineers to be able to read and maintain the code. Therefore, we, as professionals, are the only ones who can judge this. Think about it; as developers, we spend much more time reading code than actually writing it. Every time we want to make a change or add a new feature, we first have to read all the surroundings of the code we have to modify or extend. The language (Python), is what we use to communicate among ourselves.»

Citation is from [4]:

(7) «A major misunderstanding is to confuse clean code with something that can be called "beautiful code." Clean code has no beauty reasons. Professional programmers are not paid for writing beautiful or pretty code. They are hired by development companies as experts to create customer value.

Code is clean if it can be understood and maintained easily by any team member.

Clean code is the basis for being fast. If your code is clean and test coverage is good, a change or a new function just take a few hours or a couple of days – and not weeks or months – until it is implemented, tested, and deployed.

Clean code is the foundation for sustainable software and keeps a software development project running over a long time without accumulating a large amount of technical debt. Developers must actively tend the software and ensure it stays in shape because the code is crucial for survival of a software development organization.

Clean code is also a key to make you a happier developer. It leads to a stress-free life. If your code is clean and you're feeling comfortable with it, you can keep calm in every situation, even in front of a tight project deadline.

All of the points mentioned above are true, but the key point is this: Clean code saves money! In essence, it's about economic efficiency. Each year, development organizations lose a lot of money because their code is in bad shape.»

2.2. Maintainability Guidelines

«Maintainability (how easily a system can be modified) is one characteristic of software quality. Performance (how slow or fast a system produces its output) is another» [5]. The international standard ISO/IEC 25010:2011 defines maintainability as one of eight software quality characteristics.

J. Visser presents the following maintainability guidelines [5]: (1) Write short units of code. (2) Write simple units of code. (3) Write code once. (4) Keep unit interfaces small. (5) Separate concerns in modules. (6) Couple architecture components loosely. (7) Keep architecture components balanced. (8) Keep your codebase small. (9) Automate development pipeline and tests. (10) Write clean code.

Thus, maintainability is more general than clean code. Moreover, the 1st, 2nd, and 3rd guidelines are the clean code rules in [1].

Visser describes an approach for quantifying maintainability that is more practical than the emotional definitions of "clean code" in Martin's book.

Clean code and maintainability are blurred. We define the «easily-modified program» term by four simple and explicit principles.

3. Purpose of the Study

Clean code is a vague («there is no sole or strict definition of clean code») and emotional («reads like well-written prose», «beautiful code», «is also a key to make you a happier developer») term. Read the citations above again and tell what clean code is. The citations don't make the term clearer.

Maintainability is more general term than clean code. So maintainability inherits vagueness of clean code.

Purpose of the study is to introduce a criterion and explicit principles of easily-modified program, a measure to compare easily-modified program as well.

Modifying of the program consists of two main stages: (1) program reading; (2) properly program modifying. These two stages are closely related, so the principles of writing easily-modified programs apply to both the first and the second stage.

We analyze the citations, own programming experience, and formulate a criterion and principles.

4. Criterion of Easily-Modified Program

The criterion of easily-modified program is «program structure should be such as to anticipate its adaptation and modification» [6].

5. Principles of Easily-Modified Program

5.1. The names of the elements should reflect their purpose in the program

This is the cornerstone of writing easily-modified programs. Rejecting it negates all other principles.

Before changing a program, you need to understand how it works. The time to understand the work can be reduced if the program itself tells about itself, and not using the program documentation.

For the program to talk about itself, you need to name the program elements (for example, variables, routines, classes, methods, etc.), based on their purposes.

5.2. Use indentation for nested statements and blank lines

Correct naming is not the only thing that shortens the time it takes to understand the program text. Equally important is the location of the statements in a line. The indentation of statement nesting indicates the beginning and end of conditional statements, iteration statements, methods, and classes. Blank lines separate fields, constructors, methods and their logically separate parts. Avoiding indents and blank lines increases the time it takes to understand the program.

We don't mention the «only one statement in the line» rule in this section, because we suppose the rule is accepted.

5.3. One program element must have one assignment and one assignment must correspond to only one element

5.3.1. One class should do one task and one task should be done by one class

All classes can be divided into data classes and utility classes depending on their functional purpose.

The data class describes an entity of the domain. It usually contains fields and methods to get and set them values. Fields are characteristics of the entity and constitute the state of the object. Based on the principle under consideration, a data class should describe one entity, but describe it in its entirety.

The utility class has no non-const fields and only contains methods. Based on the principle under consideration, a utility class should contain methods that perform logically related actions. All other methods should be moved to other utility classes.

5.3.2. One method must do one action and one action must be done by one method

Methods create a new language that raises the level of abstraction. The program can be written in a new language using ready-made methods. If a method performs multiple actions, then it must be split. The method should be considered for splitting if its length exceeds 6-8 lines.

5.3.3. Eliminate duplicate program fragments

Duplication of program fragments is the presence in the program of a fragment that is functionally equivalent to another fragment of the same program.

There are two types of program text duplication [7]: (1) Identical: identical lines in program text. (2) Reducible: program lines can be altered to another one by such as a change of line sequence, renaming of variable, type or routine names, replacing statements by functionally equivalent ones, etc.

Duplication forces, when changing one fragment, to change its duplicate as well, which increases the time of program change.

5.4. Optimize dependencies between software elements

5.4.1. Dependency concept

The program is divided into parts, and these parts interact with each other.

A dependency is a reference in the text of one program element to another element. The program is woven from dependencies. In a class, there may be a mention of another class, for example, when inheriting or when declaring fields. In a method, methods of the same or other classes can be called.

Dependencies lead to the fact that when you change an element, you need to change all its references in other elements.

5.4.2. Optimizing dependencies

The purpose of dependency optimization is to reduce program change time. Optimization is carried out in two directions.

I. Reduction of redundant dependencies. The fewer dependencies in the program, the less time it takes to change the program. It is impossible to make software elements completely independent from each other. But some dependencies may be redundant. Therefore, they should be reduced. An easily

IOP Conf. Series: Materials Science and Engineering 1027 (2021) 012025 doi:10.1088/1757-899X/1027/1/012025

modified program has no redundant dependencies. In design patterns [8] number of dependencies is minimal.

II. Replacing some dependencies with other dependencies so as to reduce the time to change the program. Dependencies have two main characteristics:

(1) stability — the ability to remain unchanged when changing the program; the higher the abstraction used in the dependency has, the more stable this dependency is;

(2) reliability — the ability to support the performance of specified functions by associated program elements with different input data. For instance, a dependency of string and literal in line 4 is more reliable than a dependency in line 3 (Listing 1), because the dependency in line 4 doesn't throw a NullPointerException.

Listing 1. Two Approaches to Compare String and Literal

```
1  String word = "abc";
2
3  boolean isStringEqualToLiteral = word.equals("xyz");
4  boolean isLiteralEqualToString = "xyz".equals(word);
```

5.4.3. Divide programs into levels of abstraction

Program actions can be divided into levels. Level properties:

(1) actions interact only with actions of same level through data;

(2) the actions of the levels, except for the lowest one, are a sequence of actions of the immediately lower level, and the actions of the lowest level are a sequence of programming language commands.

See levels of abstraction in details in [9, 10].

5.4.4. Use as high a level of abstraction as possible

There are several aspects to this rule.

One of the abstractions is the formal parameters of methods and constructors. Using formal parameters instead of fields and constants reduces dependencies, since changing fields and constants does not entail changing formal parameters.

Using fields or constants in a method creates the following dependencies:

- to call a method with different values of fields or constants, you must first change these fields or constants, and only then call the method;
- when deleting a field or constant, you will have to change the method.

Another abstraction is abstract classes and interfaces. It is necessary to create dependencies on the most abstract elements (abstract classes, interfaces). The more abstract an element is, the more general entity it describes, and therefore changes less often. If an element changes less often, then all elements dependent on it will also change less often due to changes in this element.

5.4.5. Use dependency injection

Using static objects creates many dependencies on them in different parts of the program, which makes it difficult to modify the program. In this case a static object is like a global variable. Therefore, the use of such objects is undesirable practice.

Dependency injection allows you to eliminate this shortcoming. Dependency injection consists of two stages:

(1) creating abstraction;

(2) passing the object of this abstraction to a constructor or methods to use instead of static methods.

The abstraction is an interface or an abstract class.

IOP Conf. Series: Materials Science and Engineering 1027 (2021) 012025 doi:10.1088/1757-899X/1027/1/012025

6. Metric for Easily-Modified Programs

To declare that one program is better than another on a certain criterion, a measure must be introduced. There are four main measures of programs:

(1) productive: the time to solve the problem and the amount of memory required for this;

(2) functional: the number of functions performed by the program;

(3) complex: complexity of control and data flows;

(4) descriptive (program text parameters): number of program text lines; number of different kinds of statements (control, object-oriented, etc.) in program.

The source lines of code (SLOC) metric is not suitable for comparing easily-modified programs, since this one depends on the programming style and takes into account insignificant parts of the program.

We introduce an NOC (Number Of Commands) metric. NOC is the count of commands in class methods and static blocks. Commands are:

- calls of constructors or methods;
- arithmetic and logical operators;
- headers of the conditional statement (if) and the iteration statements (for, while, do-while);
- the return, break, and continue statements;
- the try statement.

Let's count the NOC metric for a class (Listing 2 (Listing 4–6 [1])). NOC is counted in the third column, SLOC is counted in the first column.

Listing 2. Class for Counting the NOC Metric

```
1   public class wc {
2     public static void main(String[] args) {
3       BufferedReader in = new BufferedReader(new InputStreamReader(System.in));   1 2
4       String line;
5       int lineCount = 0;
6       int charCount = 0;
7       int wordCount = 0;
8       try {                                                                         3
9         while ((line = in.readLine()) != null) {                                    4 5 6
10          lineCount++;                                                              7
11          charCount += line.length();                                               8 9
12          String words[] = line.split("\\W");                                       10
13          wordCount += words.length;                                                11
14        } //while
15        System.out.println("wordCount = " + wordCount);                             12 13
16        System.out.println("lineCount = " + lineCount);                             14 15
17        System.out.println("charCount = " + charCount);                             16 17
18      } // try
19      catch (IOException e) {
20        System.err.println("Error:" + e.getMessage());                              18 19 20
21      } //catch
22    } //main
23  }
```

The NOC metric eliminates the shortcomings of the SLOC metric. However, calculating the NOC metric is more time consuming than calculating the SLOC metric.

The NOC metric can be used in conjunction with the number of methods (NOM) [11].

The smaller the ratio of the NOC / NOM metrics, the shorter the methods and the easier the program is to change. Large NOC / NOM ratio may indicate that method is performing more than one action.

Reducing the ratio of NOC / NOM metrics should not be an end in itself. The metric values are just one of the indicators of an easily-modifyed program. In medicine, normal body temperature is not an absolute indicator of human health. Diagnostics is carried out in conjunction with other indicators, including quality ones. It's the same with programming. Metrics should be used in conjunction with other indicators, most often subjective, for example, the estimate of the program style by other programmers.

The NOC and NOM metrics are descriptive measures of program.

7. Conclusions

We have made the following conclusions.

I. Existing approaches describe clean code vaguely and emotionally. Therefore, it is necessary to formulate simple and clear criteria and principles of easily-modified programs.

II. The criterion of easily-modified program is «program structure should be such as to anticipate its adaptation and modification» from the Dijkstra's chapter of the book [6].

III. Easily-Modified program:

- has elements whose naming reflects their purpose in the program;
- is indented for nested operators and blank lines;
- has elements that have one purpose; one assignment matches only one element;
- has the minimum possible number of dependencies.

The criterion and principles are well-known. However they indeed define easily-modified programs.

IV. The number of commands (NOC) metric can be used to compare easily-modified programs.

The ease of modifying the program is a competitive advantage and an indicator of the professionalism of the programmer.

References

[1] Martin R 2009 *Clean Code. A Handbook of Agile Software Craftsmanship* (New York: Prentice Hall)
[2] Dāsa R 2016 *Learn CakePHP With Unit Testing* (New York: Apress)
[3] Anaya M 2018 *Clean Code in Python* (Birmingham – Mumbai: Packt)
[4] Roth S 2017 *Clean C++: Sustainable Software Development Patterns and Best Practices with C++* (New York: Apress)
[5] Visser J 2016 *Building Maintainable Software: Ten Guidelines for Future-Proof Code* (Sebastopol: O'Reilly)
[6] Dahl O-J, Dijkstra E W and Hoare C A R 1972 Structured programming (London: Academic Press)
[7] Avacheva T, Prutzkow A 2020 *Workshop on Materials and Engineering in Aeronautics, IOP Conf. Series: Materials Science and Engineering* **714** (1) 012001
[8] Gamma E, Helm R, Johnson R and Vlissides J 1995 *Design Patterns Elements of Reusable Object-Oriented Software* (Reading: Addison-Wesley)
[9] Dijkstra E W 1968 *Communications of the ACM* **11** 341–6
[10] Liskov B H 1972 *Proc. of Fall Joint Computer Conf.* p 191
[11] Henderson-Sellers B 1996 *Object-oriented Metrics: Measures of Complexity* (Upper Saddle River: Prentice-Hall)

IOP Conf. Series: Materials Science and Engineering 1027 (2021) 012026 doi:10.1088/1757-899X/1027/1/012026

Fuzzy joint optimization of the design and operating modes of the induction heater

N G Rogachev[1,2] and G N Rogachev[1]

[1] Automation and Information Technology Institute, Samara State Technical University, 244 Molodogvardeyskaya, Samara 443100, Russian Federation
[2] SSI SCHÄFER SAMARA, 157 Galaktionovskaya, Samara 443001, Russian Federation

grogachev@mail.ru

Abstract. This article presents an approach to improving induction heating efficiency and discusses methods for maximizing it. The problem is reduced to a fuzzy optimal calculation of the parameters of the induction heater and the heater control program.

1. Introduction

Induction heating has many advantages: low inertia, energy saving, small installation size, etc. But it also has drawbacks, the main of which is the uneven distribution of temperature over the volume of the heated body [1]. In this regard, there is a need to develop designs and operating modes of the inductor, providing the required quality of heating. The goal of a significant number of studies was to find optimal or near optimal solutions. As the manufacturing industry becomes more competitive, lowering operating costs in order to maximize profits becomes a major production goal. Since induction heating is expensive and energy intensive, lowering operating costs is a priority. Understanding how you can improve induction heating efficiency is critical to reducing energy consumption and costs.

The purpose of our study was to determine the parameters of the heater and the heating mode for the process of induction heating of paramagnetic thin-walled workpieces. The system should be close to optimal in terms of a set of specific criteria formulated in a fuzzy form. In article, this problem is reduced to the problem of a fuzzy optimal calculation of the parameters and program of operation of an induction heater.

2. Induction Heating Unit Description

The considered induction heating unit is shown in figure 1. The piece of pipe is placed inside the induction heater, consisting of an inductor powered by an AC source, and thermal insulation. The final temperature field should be fairly uniform, $T(r,l,t_{\text{FIN}}) \in [T_{\text{FIN}} - \varepsilon, T_{\text{FIN}} + \varepsilon]$.

The main control channel is the power source of the inductor $U \in [0, U_{\text{MAX}}]$. This channel allows you to influence the average temperature of the pipe, but it is not able to significantly change the law of temperature distribution along its length. The necessary uniformity of heating can be ensured by

IOP Conf. Series: Materials Science and Engineering 1027 (2021) 012026 doi:10.1088/1757-899X/1027/1/012026

choosing the length of the inductor coil, the required efficiency — by choosing the radius of the coil. An induction heater as a subject of search for an optimal design is schematically shown in figure 2.

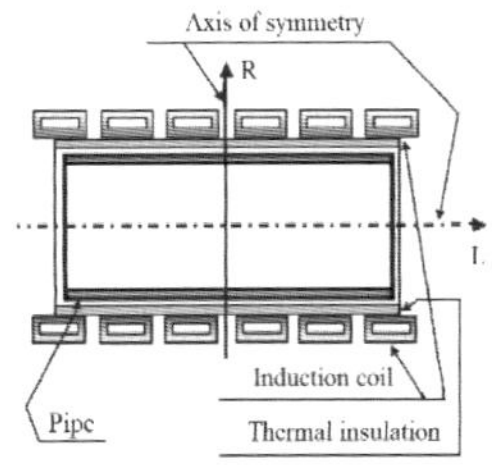

Figure 1. Induction heater drawing.

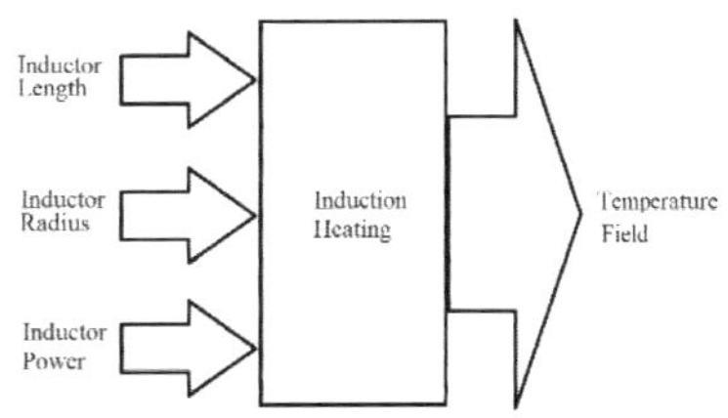

Figure 2. Induction heater as a subject of optimal design.

3. Modelling of the Induction Heating Processes

3.1. Thermal model
The thermal process in such thin-walled products as a pipe can be described by a one-dimensional heat conduction equation

$$c\gamma \frac{\partial T}{\partial t} = \lambda \frac{\partial^2 T}{\partial l^2} + p(l,t) \tag{1}$$

with boundary conditions

$$T(l,0) = T_0(l), \quad \left.\frac{\partial T}{\partial l}\right|_{l=-L_1} = \left.\frac{\partial T}{\partial l}\right|_{l=+L_1} = q(T), \tag{2}$$

where $T(l,t)$ — the temperature field of the heated product, depending on time t and spatial coordinate $l \in [0, L_1]$, $p(l,t)$ is a power of heat sources per unit volume. The temperature difference along the coordinate R of thermally thin bodies can be neglected.

Obtaining an analytical solution to the heat model (1)–(2) with an arbitrary nature of the change in the power of internal heat sources $p(l,t)$ in a visible form, convenient for further use, is impossible even in the one-dimensional case. The differential-difference method [2] is used in the work as the main method for calculating thermal fields. This method is very effective for solving such equations. In this method, the partial derivative with respect to the spatial coordinate is replaced by a finite difference. This allows the partial differential model (1)–(2) to be reduced to the ODE system

$$\frac{d\overline{T}}{dt} = A\overline{T} + B\overline{p} \tag{3}$$

for the sets of temperatures $T_i(t)$ and power of heat sources $p_i(t)$ per unit volume at the sampling points l_i.

However, the use of standard programs for calculating thermal fields for fuzzy optimization of induction heating processes seems inappropriate due to the low efficiency of such programs under these conditions, which do not take into account the specifics of the processes and tasks under consideration. Therefore, the authors have developed their own routines [3].

3.2. Electromagnetic model
The volumetric power density of internal heat sources $p(l,t)$ characterizes the intensity of internal heat release that occurs in a heated product during induction heating. Calculation of the distribution of internal heat sources is a complex independent task and is carried out during the simulation of electromagnetic processes [4].

As a method of modeling electromagnetic processes during induction heating, we use a numerical method known as the secondary source method [5]. The calculation of fields in non-magnetic bodies according to it reduced to determining the current $\dot{I}_N$ in the inductor and currents $\dot{I}_i$, $i = 1, 2, \ldots, N-1$ in $N-1$ elementary solenoids into the body. All of these solenoids are magnetically coupled to each other (figure 3). To find the currents, a system of equations of the form

$$[\dot{Z}]\cdot[\dot{I}] = [\dot{U}] \tag{4}$$

obtained by applying the 2nd Kirchhoff law to each of the solenoids. In (3) the total impedance $[\dot{Z}] = [R] + j[X]$, where $U_i = \begin{cases} 0, & i \neq N \\ U, & i = N \end{cases}$; $R_{ik} = \begin{cases} R_i, & i = k \\ 0, & i \neq k \end{cases}$, R_i is the active resistance of the i-th solenoid, $X_{ik} = \begin{cases} \omega L_i, & i = k \\ \omega M_{ik}, & i \neq k \end{cases}$, L_i — inductance of the i-th solenoid, M_{ik} — mutual inductance of the i-th and k-th solenoids, $\omega = 2\pi \cdot F$ is the angular frequency.

The solution of (3) determines the value of the currents in the inductor and in the solenoids, which represent the components of the body's electromagnetic system. Solenoid currents are used below to determine the power of heat sources p_i in (3),

$$p_i = I_i^2 \cdot R_i. \tag{5}$$

For the numerical simulation of electromagnetic processes of induction heating, the authors developed special subprograms, which, together with subprograms for calculating thermal processes, comprise a digital model of induction heating of paramagnetic thin-walled shells. The model is implemented in the MATLAB and was used for fuzzy optimization of a heating process. Some results of calculating the power density of heat sources are shown in figure 4.

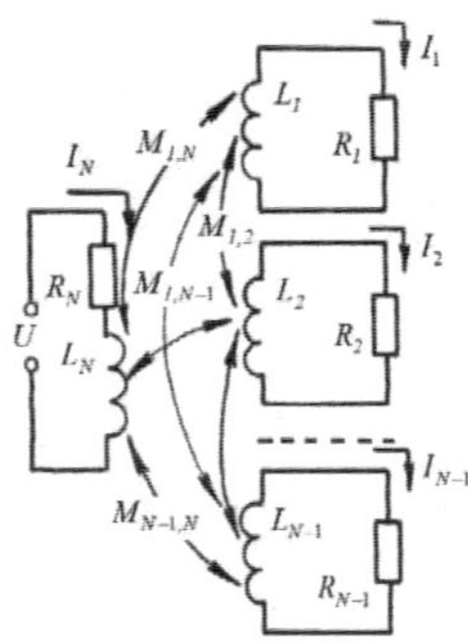

Figure 3. Equivalent circuit of an induction heater in the form of several magnetically coupled circuits.

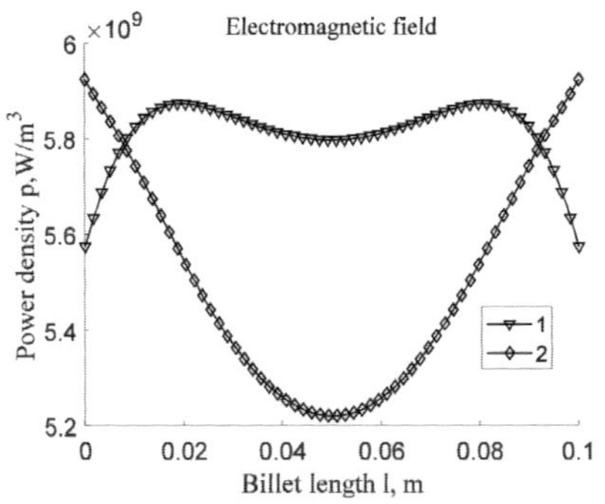

Figure 4. Distribution of the power density of heat sources p along the length l of the heated product. 1 — inductor radius $R_2 = 0.055\ m$, inductor length $L_2 = 0.140\ m$; 2 — inductor radius $R_2 = 0.058\ m$, inductor length $L_2 = 0.165\ m$.

4. The main quality indicators of an induction heating

The main indicators of the quality of an induction heater are its net power P_N, total power P_T, efficiency $\eta = (P_N / P_T)\cdot 100\%$, time t_{FIN} and uniformity of billet heating ε. The uniformity of the electromagnetic field during the heating process, as well as the uniformity of the temperature field at the end of the process, can be estimated by the inf-norm $\|\cdot\|_\infty$. The influence of the parameters of the induction heater on these criteria is analyzed using the electromagnetic model and thermal model considered in the [3].

5. Numerical example

This part is devoted to the presentation of some numerical results. Here is an example of simulating a long inductor (the size of a workpiece) for heating a paramagnetic thin-walled pipe.

The numerical results presented here allow a better understanding of the behavior of temperature and electric field depending on the parameters of the inductor. Moreover, these results motivate the statement of the fuzzy optimization problem. The calculations were performed on a Core i5 processor in MATLAB.

We simulated the process of induction heating up to a temperature $T_{\text{FIN}} = 900\ K$ of a pipe with a heat capacity $c = 500\ J/K$, density $\gamma = 7800\ \text{kg}/m^3$, thermal conductivity $\lambda = 20\ W/(m \cdot K)$, resistivity $\rho = 2 \cdot 10^{-6}\,\Omega \cdot m$, outer radius $R_{1O} = 0.05\ m$, inner radius $R_{1I} = 0.04\ m$, length $L_1 = 0.1\ m$. The parameters of the induction heating installation: the number of turns w = 12, the frequency and power of the power source — 2400 Hz and 400 kW, the radius R_2 and length L_2 of the inductor will be determined during design.

The simulation results are presented in the figures 5–8 as dependences of the main indicators of the quality of induction heating and operating mode U on the parameters R_2 and L_2 of the inductor. The given graphs demonstrate multidirectional tendencies depending on the quality characteristics of the object on its parameters and operating mode. Indeed, the electrical efficiency decreases with an increase in the length of the inductor and its radius, the thermal efficiency increases with an increase in the radius of the inductor and decreases slightly with increasing length. This leads to the fact that the overall efficiency decreases with increasing length of the inductor and increases with increasing radius of the inductor. The best uniformity of the electromagnetic field is achieved when the parameters and operating conditions of the installation differ from those at which the best uniformity of the temperature field is ensured.

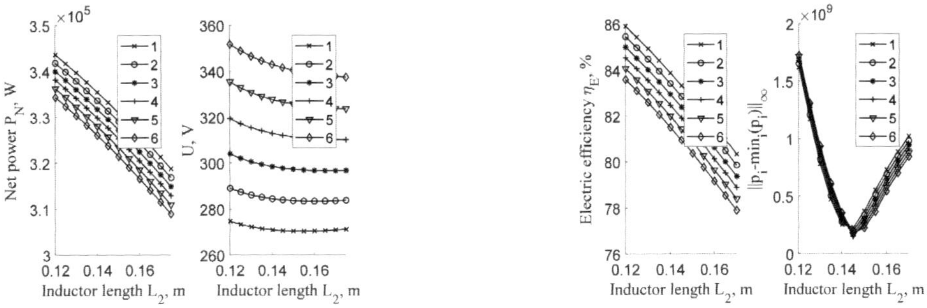

Figure 5. Dependences of the main indicators of the quality of induction heating and operating mode U on the length L_2 of the inductor. 1 — $R_2 = 0.053\ m$; 2 — $R_2 = 0.054\ m$; 3 — $R_2 = 0.055\ m$; 4 — $R_2 = 0.056\ m$; 5 — $R_2 = 0.057\ m$; 6 — $R_2 = 0.058\ m$.

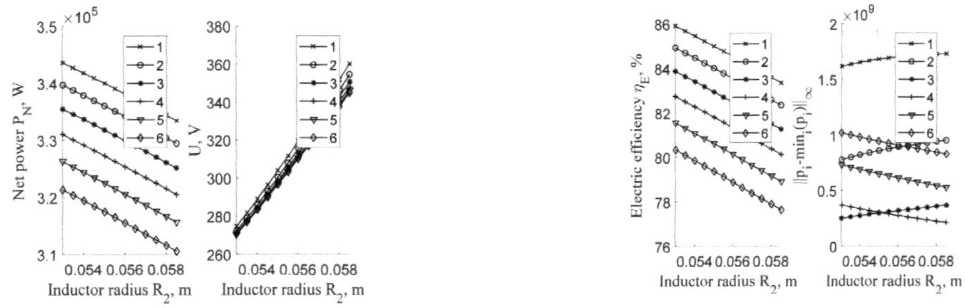

Figure 6. Dependences of the main indicators of the quality of induction heating and operating mode U on the radius R_2 of the inductor. 1 — $L_2 = 0.12\ m$; 2 — $L_2 = 0.13\ m$; 3 — $L_2 = 0.14\ m$; 4 — $L_2 = 0.15\ m$; 5 — $L_2 = 0.16\ m$; 6 — $L_2 = 0.17\ m$.

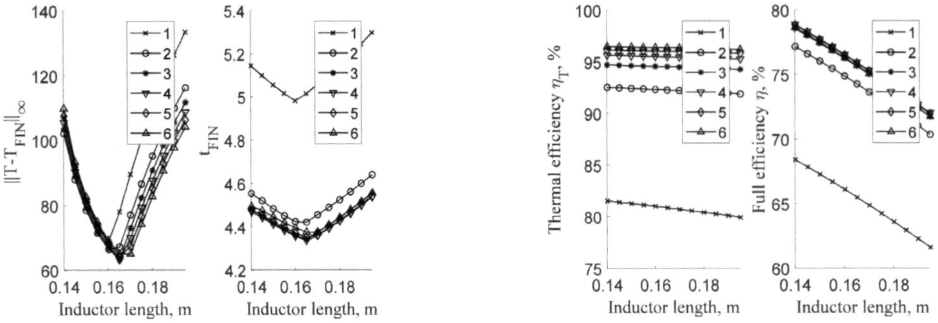

Figure 7. Dependences of the main indicators of the quality of induction heating and operating mode U on the length L_2 of the inductor. 1 — $R_2 = 0.053\ m$; 2 — $R_2 = 0.054\ m$; 3 — $R_2 = 0.055\ m$; 4 — $R_2 = 0.056\ m$; 5 — $R_2 = 0.057\ m$; 6 — $R_2 = 0.058\ m$.

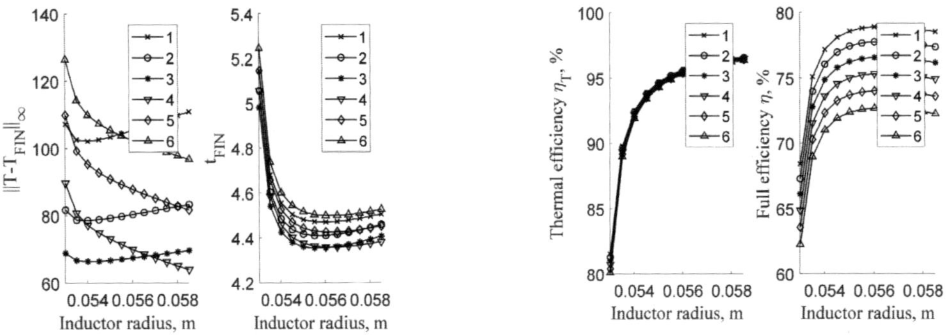

Figure 8. Dependences of the main indicators of the quality of induction heating and operating mode U on the radius R_2 of the inductor. 1 — $L_2 = 0.14\ m$; 2 — $L_2 = 0.15\ m$; 3 — $L_2 = 0.16\ m$; 4 — $L_2 = 0.17\ m$; 5 — $L_2 = 0.18\ m$; 6 — $L_2 = 0.19\ m$.

Based on this, the problem of determining the best combination of parameters and operating mode of the heater was formulated as a fuzzy optimization problem [6, 7].

6. Statement of the problem of joint fuzzy optimization of the design and operating modes of an induction heater

In verbal form, the technological requirements for induction heating look like this. The product must be heated to the temperature $T_{FIN} = 900\ K$ in the minimum time $t_{FIN} \to \min$, the temperature drop along the length should be minimal, $\varepsilon \to \min$. For maximum efficiency of induction heating, a verbal description of the goals and constraints will be as follows: heating must be carried out with maximum efficiency $\eta \to \max$, with enforcement of high-frequency power supply constraint $P_T \le 400$ kW.

The design problem with fuzzy goals and constraints is formulated as follows. It is necessary to transfer model (3)–(4) during time $t \in [0, t_{FIN}]$, $t_{FIN} \to \min$ to point $\max_{l \in [0, L_2]} |T(l, t_{FIN})| \le \varepsilon$ under additional condition $\varepsilon \to \min$. Further, it is necessary that the conditions $\eta \to \max$, $P_T \le 400$ kW are satisfied all the time $t \in [0, t_{FIN}]$.

Sigmoidal membership functions of fuzzy sets are defined analytically: $\mu_1(\varepsilon) = 1 - (1 + \exp(0.1 \cdot (70 - \max_{l \in [0, L_2]} |T(l, t_{FIN})|)))^{-1}$; $\mu_2(t_{FIN}) = 1 - (1 + \exp(5 - t_{FIN}))^{-1}$;
$\mu_3(\eta) = 1 + \exp(0.1 \cdot (70 - \eta))^{-1}$.

Further, the problem is reduced to calculating such parameters L_2, R_2 and operating mode U of the heater, which together provide the maximum possible value of the minimum of all membership functions $\mu_i(\cdot)$, $i = 1, 2, 3$, i.e.

$$\max_{L_2, R_2, U} \min_i(\mu_i),\ i = 1,2,3. \tag{6}$$

7. Solving the problem of joint fuzzy optimization of the design and operating modes of an induction heater

Problem (6) is solved numerically. As a procedure for the numerical solution of problem (6), the MATLAB function fminimax [8] was used, which allows minimizing the maximum values of the set of objective functions, taking into account linear and nonlinear constraints.

The characteristics of the found solution are $U = 318$ V, $L_2 = 0.1623$ m, $R_2 = 0.0566$ m. Results are $\varepsilon = 63\ K$, $t_{\text{FIN}} = 4.25$ s, $\eta = 76$ %. As shown in figures 9–11, the degree of satisfaction of the fuzzy requirements is 0.65.

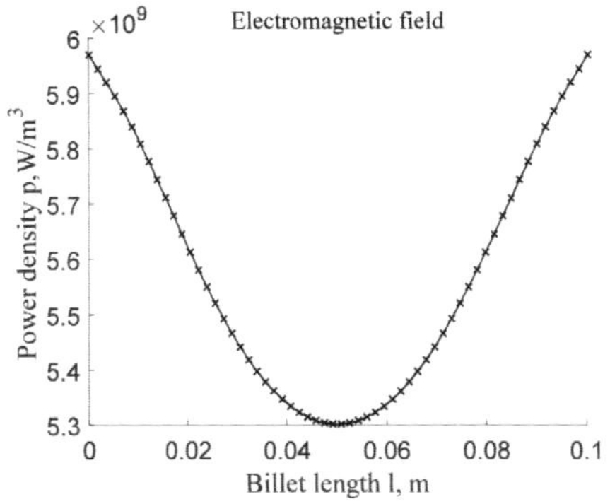

Figure 9. Distribution of the electromagnetic field along the billet length.

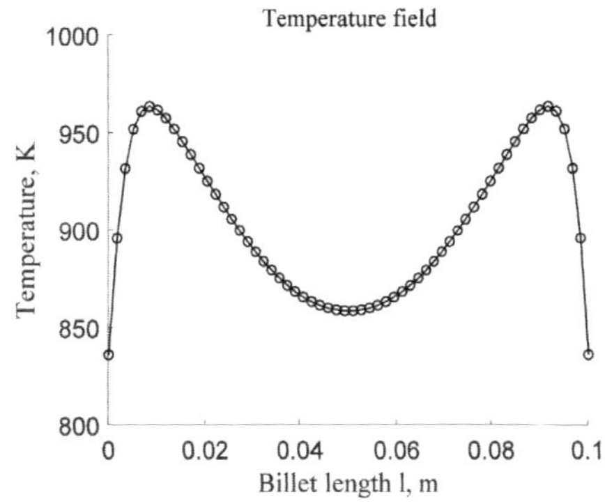

Figure 10. Distribution of the temperature field along the billet length.

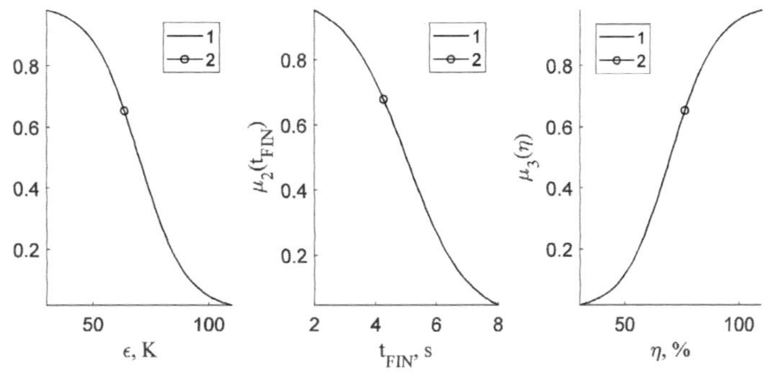

Figure 11. Results of fuzzy optimization of the design and operating mode of the induction heater. 1 — Membership functions of fuzzy sets; 2 — Membership of the calculated fuzzy-optimal solution.

8. Further research

Further research will focus on calculations in the task of a multi-stage technological process, including heating and subsequent transportation of billets to equipment for subsequent processing.

Acknowledgments
The reported study was funded by RFBR, projects number 18-08-00506 and number 19-38-90061.

References

[1] Rudnev V, Loveless D and Cook R L 2017 *Handbook of Induction Heating* (Boca Raton: Taylor & Francis)
[2] Rapoport E Ya 2005 *Analysis and Synthesis of Automatic Control Systems with Distributed Parameters* (Moskow: Higher School)
[3] Rogachev G and Rogachev N 2019 Modeling and fuzzy optimization of thin-wall shells induction heating *Proc. XXI Int. Conf. Complex Systems: Control and Modeling Problems (Electronic Materials)* (New-York: Institute of Electrical and Electronics Engineers Inc.) pp 245–248
[4] Sluhotsky A E and Ryskin S E 1974 *Inductors for Induction Heating* (Leningrad: Energy)
[5] Nemkov V S and Demidovich V B 1988 *Theory and Calculation of Induction Heating Devices* (Leningrad: Energoatomizdat)
[6] Bellman R E and Zadeh L A 1970 Decision-making in a fuzzy environment *Management Science* **17** 4 141–64
[7] Carlsson C, Fullér R and Giove S 1998 Optimization under fuzzy rule constraints *Belgian Journal of Operations Research, Statistics and Computer Science* **38** 17–24
[8] Yang W Y, Cao W, Chung T-S and Morris J 2005 *Applied Numerical Methods Using MATLAB* (Hoboken: John Wiley & Sons)

IOP Conf. Series: Materials Science and Engineering 1027 (2021) 012027 doi:10.1088/1757-899X/1027/1/012027

Adaptive imaging system for electromagnetic scattering field of aircraft

A T Tarlanov and Z M Kurbanismailov

MIREA – Russian Technological University, Moscow 119454, Russia

lev.brave@gmail.com

Abstract. Electromagnetic compatibility (EMS) of a sophisticated aircraft technical system with other systems is a prerequisite for their sustainable operation. Aerospace technology, where both highly sensitive measurements and energetic control effects are closely combined in the dense layout of an aircraft, requires a solution to the EMS problem. This paper shows the need for effective technology to control the electromagnetic radiation created in the surrounding space, both by subsystems of aircraft and by the whole object. The method how to of visualize of 3D electromagnetic field both in space and in time is considered.

1. Introduction

Electronic devices have become an integral part of human life and any technical system, that has interactive properties and requires certain conditions for sustainable functioning. One of these mandatory conditions is EMS [1] of the entire system and its subsystems with each other, there are no less complex and sensitive. Although digital technology is fundamentally highly disruptive, but the modern technical system contains a number of analog sensors, communication lines, receiving antennas, which are sensitive to the external interfering signals both through wires and the air. Hence, the problem of EMS is relevance for integrated technical systems, that perform of measuring, data processing or control commands creating either digital or analog. One example is aerospace technology, where highly sensitive measurements and energetic control influences are closely combined in a dense aircraft layout. EMS of such systems are being provided at all stages of the aircraft lifetime, from development, to stand-up tests, including operational phases. EMS tools [2] are screening, time and frequency separation, minimizing of cross-influence by means of design and circuitry methods, with objective visualization of their results. To quantify the EMS of the aircraft, an effective technology is needed to control the electromagnetic radiation fields, generated by both individual subsystems and the whole object. It is shown that existing imaging methods do not provide to select the observation plane in the 3D-space of the observation zone, as well as organize a visual 3D visualization of the scattering field spatial distribution around the object. The aim of this work is to formulate proposals for the spatial distribution visualizing of physical fields for different applications. The work proposes a system of assessment of external electromagnetic radiation of an aviation, which built on mapping the radiation field in a limited observation zone on a given 3D grid [3]. It is proposed an algorithm for measurement results imaging by either brightness changing of markers, and in a given a priori color scale. The problems of in time animation, imaging of measured values and spatial visualization have been solved.

2. Background definitions

2.1. Definition of Visualization

Popular Wikipedia [4] defines the visualization procedure as a method of presenting information in a visual form, in a form, suitable for studying with the help of the human eye. Everyone knows the visualization of the magnetic field with iron sawdust, or visualization of the flow of liquid or gas by injecting them with contrasting substances (smoke in the gas, bubbles in the liquid).Visualization of the electromagnetic field in our task involves discretizing of the field in the nodes of a 3D grid with a cubic shape.

2.2. Overview of experimental data visualization techniques

The visual representation of the results has always been of interest as a final phase of any study, answering the question of the size and nature of the measured or calculated value. Oscillographs [5] have been a means of measurable values visualizing up to units of seconds duration, in the very recent past. Also the recorders [7] were widely used, which used, with record by "ink" on the paper. Visualization of fast and long processes in a wide range with simultaneous registration became possible only as a result of the computers with advanced software. Popular MatLab and MathCad computing packages have advanced imaging tools for calculating results, both in the form of 3D spatial imaging and multi-channel time dependencies. However, these packages are off-line applications, and their alignment with real-time systems causes a number of difficulties, which were not provided by these data computational tools. Advanced computer imaging programs are represented by universal products [8–10] such as Google Data Studio etc. These products are widely being used in business analytics, logistics and allow to give a visual view of the calculation data in the form of charts, functional graphs and reports. Attention is paid here to the order of information placement, the creation of visual images. That helps an analyst or manager to make the right decisions by identifying trends and calculating statistical estimates, based on the available information. However, in scientific tasks, these software products have limited application, as they do not allow to carry out spatial 3D visualization, they work in off-line mode, having no connection with the external interface, and do not display 3D processes in real time. Besides, the developer of these tools does not give Software Development Kit, which limits the possibilities of their application.

3. Software creation methodology

3.1. Determining the number and position of the settlement grid nodes

It is proposed the equidistant position of the grid nodes in the space of the cubic form observation zone, which allows to attribute the calculated or measured values of inductions of fields [11–14] to specific points of space, surrounding the study object. The coordinates of the control grid nodes are a parameter, that allows to quantify and link the electromagnetic field induction to the object coordinate system. The 3D grid from the checkpoints in the nodes fits into the observation zone. The number of dots on one side of the controlled zone is indicated by the natural digit, making it easy to change the size and detail of the visualization system in question at the software level. The number PPC points, located on an arbitrary selected section of the observation plane, as shown in Figure 1, is determined by the formula:

$$PPC = R^2, \quad (1)$$

where: R — the number of points on one of the axes of the chosen plane.

The total number of points in the specified volume can be calculated by the formula:

$$PC = R^3, \quad (2)$$

The numbering of the points begins at the bottom left rear edge of the volume with the rectangular Descartes coordinate system, i.e. at X < 0, Y < 0, Z < 0, and continues first on the X axis, then on the Y and Z with increment PS, calculated by formula:

$$PS = \frac{V}{R-1}, \tag{3}$$

where: V — the volume of the calculated cube [m³].

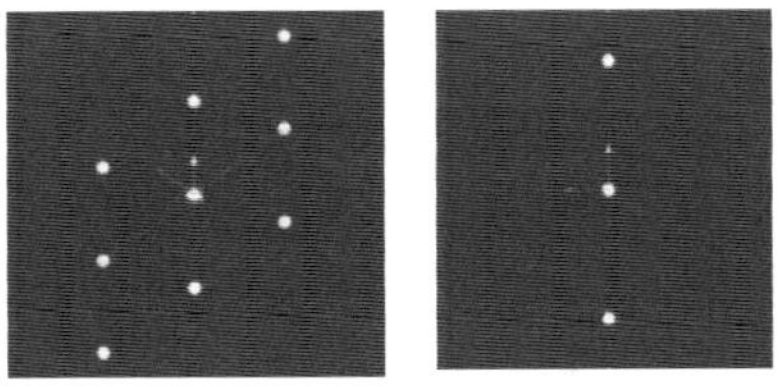

(a) (b) **Figure 1.** Point on the plane: (a) — promising view; (b) — side view.

The position of each settlement point on the X, Y and Y axes can be calculated by formulas:

$$PX = MOD(I, R) * PS - \frac{V}{2}, \tag{4}$$

$$PY = MOD\left(\frac{I}{R}, R\right) * PS - \frac{V}{2}, \tag{5}$$

$$PZ = \frac{I}{PPC} * PS - \frac{V}{2}, \tag{6}$$

where: *MOD* — the result of dividing with the balance of the first argument for the second, *I* — the index (number) of the settlement point. This technique is operating equally well in both the right-hand and left-hand coordinate system, but we will use a traditional left-handed system in this paper. Using formulas (1–6) it can generate a 3D cube of settlement points, as shown in Figure 2 (a), which will link the results of measurements to a specific system of coordinates of the observation object. The numbering of the calculation points of the cubic settlement grid with a resolution of 3 points on the side is shown in Figure 2 (b). The cube grid, which contains 7 points on each side, is shown in Figure 2 (c). Measurements in the nodes of the calculation grid with the above coordinates allow to create an measured data array reflecting the scattering field of the object at the time of observation. The size of the array is being calculated by formula (2), and the resolution and volume of the cubic observation zone is easy to control by changing the R and V parameter accordingly. The above formulas allow you to select the volume and resolution of the observation zone needed to solve problems, depending on the dimensions of the observed object. At the same time, each node of the calculated grid is attributed to the module value of the induction vector of the measured scattering field, which is coming from the measuring system. An example of the implementation of a qualitative picture of the field with a black-white scale of the observed field induction is shown in Figure 3.

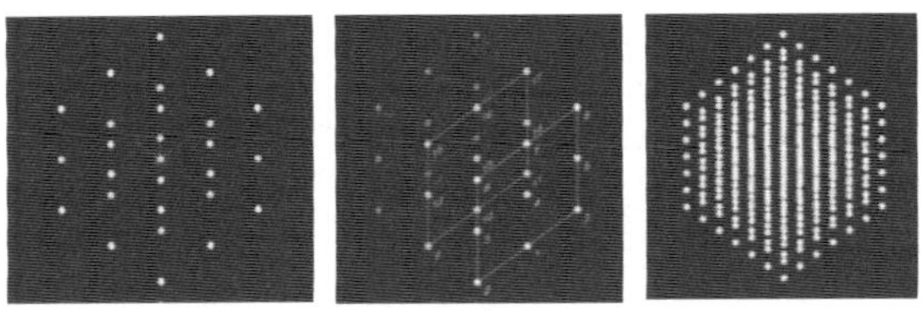

(a) (b) (c) **Figure 2.** The cube of the settlement grid, embedded into the observation zone of the scattering fields, in this case with a resolution of 3 points on the side of the cube (a). Indexing (numbering) of the observation zone settlement points (b). Cubic settlement grid with a resolution of 7 points on the side (c).

Here the dark dots correspond to the weak field, white — strong one. The scale of induction is proportional to the brightness of the marker and extends from black to white. Visualization of such a black-and-white map provides an opportunity to get a qualitative view of the electric or magnetic field induction into the observation area, which can be very useful in the first stages of the study.

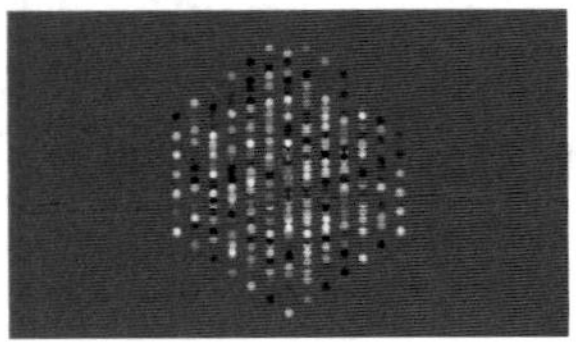

Figure 3. A point view of field induction with proportional adjustment of the markers brightness in mesh nodes.

Further processing of the presented results allows to interpolate the values of induction in the nodes of the grid — to arbitrary intermediate points in the space between the nodes. Moreover, interpolation can be carried out on the surface of the conventional plane, arbitrarily located to the determined axes of the coordinate system, which allows to carry out detailed visualization in the chosen directions.

3.2. Calculating the settlement points, which include the estimated pixel

The minimally described geometric plane consists of three dots forming it. This plane is used in computer graphics to represent the mesh. A mesh is a set of points in space and their connections to each other, forming a certain number of surfaces, sometimes with adjacent edges. The mesh belongs to the category of vector graphics, which makes it necessary to roster it before putting it on display. Rostering [16] is the process of converting data from a 3D view to a 2D image on a selected surface, consisting of individual pixels. The coordinates of each pixel of the plane are calculated from the coordinates of the nearest fixed grid nodes described above. This pixel will be called as the estimated pixel. The estimated pixel stores information about its color (the value of the induction) and the position in the 3D space of the grid. This information is used to determine the neighboring indexes of offset points between which the pixel was located. A formula is used to calculate a smaller offset index:

$$RL = FLOOR\left(\frac{PP + \frac{V}{2}}{PS}\right), \tag{7}$$

where: *FLOOR* — rounding function down to the nearest whole digit, PP — position on a specific axis of the calculated pixel in space. As a *PP*, its need to position the pixel on each axle by determining the shift index on that axis, which the estimated pixel is to the right, above or deeper, depending on the axis that is framed. The index obtained in Formula 7 can go beyond the calculated volume. To prevent this let's limit the value of offset indices to the maximum threshold (cube boundary):

$$IRL = MIN(RL, R-1), \tag{8}$$

$$IRH = MIN(RL+1, R-1), \tag{9}$$

where: *MIN* is a function of finding among at least two parameters, RL is a smaller offset index without taking into account of going beyond. Formulas (7–9) determine the indexes of offsets of settlement points horizontally (axis X), vertical axis (Y) and depth (axis) inside or on the surface of which is the estimated pixel situated. This is necessary for the bi-linear interpolation of the estimated pixel value between the surrounding points in the observation zone space.

3.3. Interpolation of a pixel within a grid element

In the next step, its necessary to calculate the position of the pixel relative to the surrounding 8 points of the nearest grid nodes. To do this, let's first calculate the position of the pixel relative to the coordinates of any of the surrounded points of the nodes. In this technique, the left lower rear settlement point (I1) of 8 volumes is selected for this, as shown in Figure 4. In the figure, the conditional settlement pixel (white sphere) is located in the geometric center of the nodes points and its coordinates are counted (red arrow) from the initial settlement point of the I1 node. To calculate a normalized position, the pixel coordinates are being divided onto the distance between the dots — the step of the grid. Based on the conditional equitability of the calculated pixel from the grid nodes, its

IOP Conf. Series: Materials Science and Engineering 1027 (2021) 012027 doi:10.1088/1757-899X/1027/1/012027

normalized coordinates in this case will be [0.5, 0.5, 0.5] which are calculated by the following formula:

$$NP = \frac{WP - I1}{PS}, \tag{10}$$

where: *WP* is a vector, representing the position of the estimated pixel in the global space. The normalized coordinates of the pixel reflect its proximity to the nodes of the settlement grid. The lower the value of the normalized coordinate on the chosen axis of coordinates, corresponds the closer the estimated pixel to the settlement point I1, the more — the closer to the next point on the same axis.

3.4. Converting offset indexes into a settlement point index

To convert normalized pixel offset indexes within the grid element — into the real coordinates of the settlement grid, the following formula is proposed:

$$I(HI, VI, DI, R) = HI + VI * R + R^2 * DI, \tag{11}$$

where: HI — horizontal shift index, VI — vertical one index, DI — depth shift index. This formula should be applied as many times as the nodes are surrounded by the pixel in question, namely n = 8, each time substituting the coordinates of the relevant node:

- $I1 = I(HL, VL, DL, R)$; $I2 = I(HL, VH, DL, R)$; $I3 = I(HH, VH, DL, R)$;
- $I4 = I(HH, VL, DL, R)$; $I5 = I(HL, VL, DH, R)$; $I6 = I(HL, VH, DH, R)$;
- $I7 = I(HH, VH, DH, R)$; $I8 = I(HH, VL, DH, R)$;

where: *HL* and *HH* — the smaller and higher value of the index horizontally, *VL* and *VH* — the smaller and greater value of the vertical shift index, *DL* and *DH* — the smaller and greater value of the index of shift in depth.

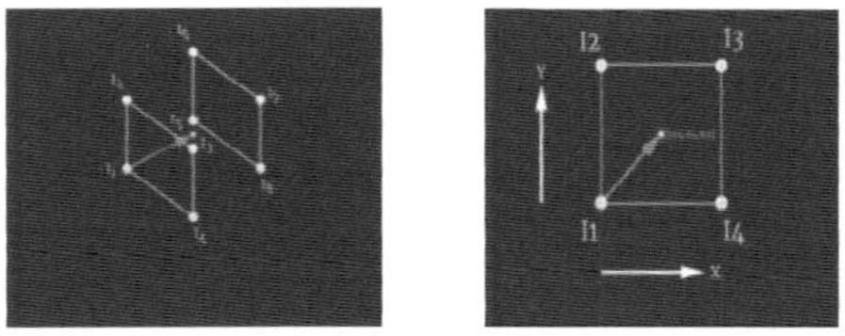

Figure 4. Relative position of the estimated pixel (perspective and side view).

3.5. Bi-linear interpolation between points

With normalized pixel coordinates within the settlement grid element, as shown above, it will interpolate the coordinates [17] between the mesh nodes opposite each other and then between the results. This method of interpolation is called bi-linear. Each pixel is assigned a vector of electric field induction, coming from the measuring system, or the vector of magnetic field induction, depending on the type of measurements on which visualization is carried out at the moment. This implies that the measuring system performs vector measurements of the electromagnetic fields. Each vector is described as the size of a module that is not oriented-dependent, and the direction which set by the orthogonal component ratio of the vector. First, its necessary to interpolate the values of the modules by the values of normalized coordinates, calculated in formula 11. The interpolation formula between the two values is as follows:

$$LI(F, T, P) = F + (T - F) * P, \tag{12}$$

where: *F* — initial value, *T* — the final value, *P* — relative shift from *F* to *T*. This formula also works with the orthogonal component of vectors by applying it to their individual components:

$$LIV = \big(LI(V1X, V2X, P), LI(V1Y, V2Y, P), LI(V1Z, V2Z, P)\big), \tag{13}$$

where: V1X, V1Y, and V1 — first vector components, V2X, V2Y, and V2 — values of the second vector components, P is a relative shift from the first vector to the second. The length of the vector is calculated by the formula:

$$L = \sqrt{X^2 + Y^2 + Z^2}\,, \tag{14}$$

where: X, Y and Y are components of the measured vector on the X, Y and Z axes respectively. Since there are 8 nodes of the settlement grid element, let's first interpolate along the vertical axis between I1 and I2, using the normalized coordinate value Y of the calculated pixel as an interpolation indicator. Then we do the same procedure between I4 and I3, and the results are also interpolated with each other by the value of normalized coordinates X of pixel as shown in figure 5 on the left. The result of this operation is a vector of electric or magnetic field induction at the point, located directly on the plane, formed by the calculation points I1, I2, I3 and I4. The same bi-linear interpolation must be done with points I5, I6, I7 and I8. The results of both planes are interpolated among themselves by the value of the normalized coordinates Z of the pixel, as shown in Figure 5 in the center and right. This calculates the final vector of electrical or magnetic field induction in pixel coordinates. By calculated values it's possible to set the brightness of the pixel, displayed in the range of black to white on the chosen plane, arbitrary crossing the specified volume of the observation zone, as shown in Figure 6. It can be seen that the brightness of the pixels reflects the measurable modular values of [11, 12], and the adjacent areas are also bi-linearly interpolated to illustrate the transitions. The created software allows to choose the arbitrary position of the control plane, on which the modular images are rendered, such as measurement results, as shown in Figure 6 (b). The image shows that the visualization plane is interpolated between the adjacent planes of the nodes. The high gradience of the background on the visualization plane is due to the small number of mesh nodes at the selected observation area interval. Increasing this number even to n = 10 allows to get a smoother distribution of the measured and interpolated image of the field, as shown in Figure 6 (c). Here, dark areas have a lower value of the measured vector module, light — large. It is shown, that the arbitrary movement, rotation, and scaling of the visualization plane within the scope of the observation — does not affect the accuracy of the interpolation and the representation of the measurable values, coming from the measuring system.

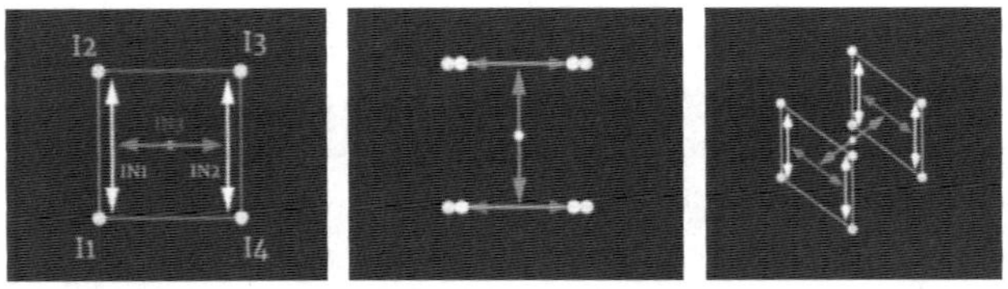

Figure 5. Bi-linear interpolation between points (left). Interpolation between interpolated plane values (top view — in the center, and from perspective).

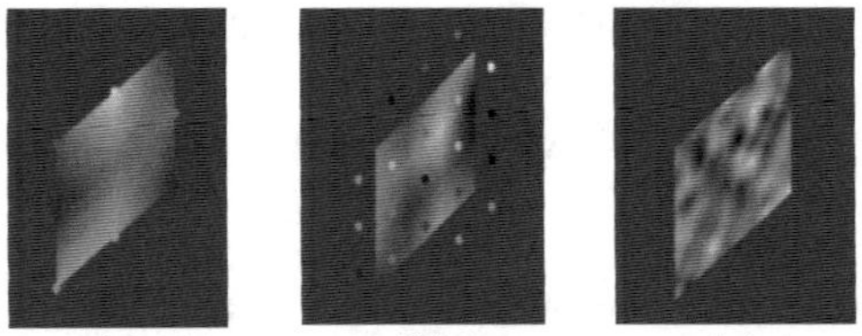

(a) (b) (c)

Figure 6. Selected plane displaying modular values of measurable values based on interpolation results (a). A visualization plane located between two planes of the settlement grid nodes (b). The plane visualized by the algorithm at higher resolution (c) — grid nodes are hidden for clarity.

3.6. Animation in Time

To time animation the visualization either modular and vector measurements, it is possible to periodically replace the array of data coming from the measuring system, which is carried out through the time interval set at the program. Since the research of the EMS is mainly carried out in a stationary mode, the intervals of replacement time of arrays are set in the range of units of seconds, which is intended to study the long-term stability of the observed radiations. However, the software has no theoretical limitations for reducing the observation time interval to a small parts of a second, if necessary.

3.7. Volume visualization of the scattering fields measurements results

For 3D imaging, it is proposed to create a series of different cubic volumes, limited by the visualization planes, described above, which are located symmetrically relative to the geometric center of the observation area. Moreover, the cubic volumes have different sizes, in a scale that changes with a constant step from the first volume to the last, for example in the total amount of the N = 100. Besides, the created cubic volumes have a transparency of 99%, which allows to observe the measured values through all the planes of the created cubic volumes of visualization, as shown in Figure 7. This mechanism, proposed in this paper, also allows to understand what the distribution of electromagnetic fields inside the observation zone, when the coordinates of the external observation point are being changed.

3.8. Color map

The proposed software allows to display the scale of the magnitude of the measured induction not only by the brightness of the marker, but also by its color. To do this, the color marking of the interpolation result from the formula (14) is proposed, in according to the scale, shown in Figure 8.

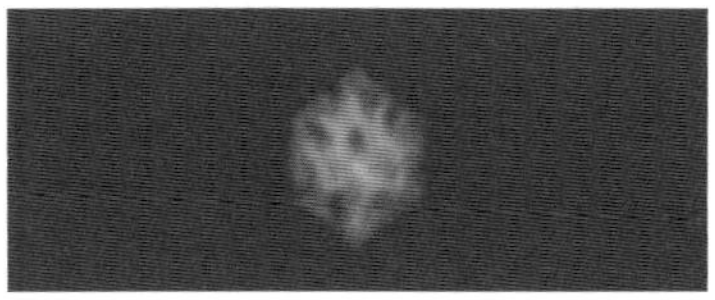

Figure 7. Visualization of measurable values in the volume of the observation area.

Figure 8. Map of matching colors and measurable values.

The higher measured value in a particular pixel corresponds the more red the shade of marker becomes, the lower the bluer. The left part of the color map corresponds to the minimum value of the measured value, the right part — the maximum. The scale boundaries of the measured values are set at the program level in the initial dialogue, together with the volume of the observation zone and the step of the calculation grid. Based on above boundary values it's offered the following expression to calculate the relative CPX pixel coordinate on the above scale, a color matching map:

$$CPX = \frac{L - MIV}{MAV - MIV}, \tag{15}$$

where: L — the length of the induction vector in the calculated pixel, MIV — the minimum value among all settlement points, MAV — the maximum value among all settlement points. Formula 15 allows you to calculate the pixel percentage position on the color map. The following expression is proposed to extract the color directly:

$$TEX = FLOOR(ABS(MOD(CPX, 1)) * W), \tag{16}$$

where: $FLOOR$ — rounding function down to the nearest whole digit, ABS — function extracting absolute value, MOD — fission function with the remainder of the first argument on the second, W — the width of the texture in the pixels. This function extracts the color of the pixel from the above color map, which is the result of the color visualization algorithm, shown in Figure 9. In the figure, the small values of the measured values (dark spheres) correspond to colder colors on the chosen visualization plane, strong values (light spheres) — hotter. As in previous cases, the visualization plane may be arbitrarily between the nodes of the settlement grid, which does not affect the correct display of the marker color, according to the chosen scale. A more detailed picture of the modules measured distribution is shown in Figure 11 (b), which is achieved by reducing the step of the settlement grid. The proposed color imaging techniques also can be apply to all animation algorithms, described above — in time, 3D, or interpolation at the arbitrary position of the chosen visualization plane.

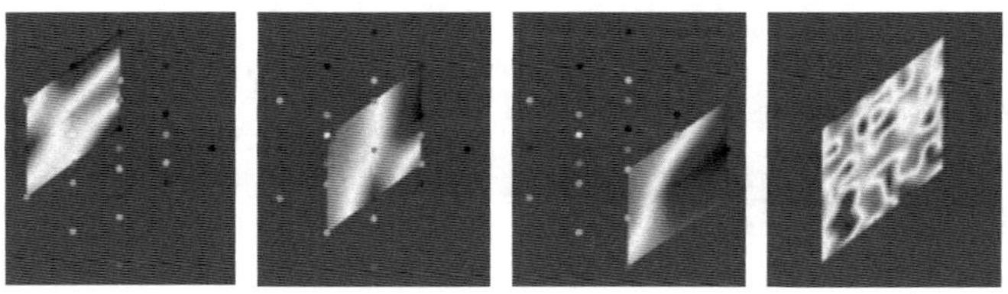

Figure 9. Color visualization of modular measurements.

4. Results

4.1. Description of the resulting product

Thus, the visualization system is part of the system of mapping the electromagnetic radiation of the object, the structural scheme of which is shown in Figure 10. The figure shows the following elements of the mapping system: P — positioner linear, ID — measuring device, INT — interface between the measuring device and software product, PPV — described in this paper software application "Visualization." The positioner is a linear manipulator that is designed to install a measuring sensor at the point of the observation zone, with the coordinates, which set in the grid by initial dialogue with the software visualization product. At the initial stage, the measuring sensor can be moved to the specified distances by some linear guide elements — rakes, oriented along the axes of a given system of coordinates of the object of the study. A measuring device is a vector magnetometer for measuring the magnetic field, or a measuring capacitor to measure the electric field. The interface is based on the PC1716 device [20], which multiples measuring channels, makes analog-digital conversion, and transfers information to the software environment in the language of C.

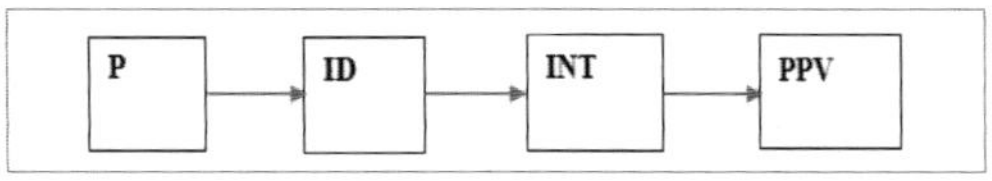

Figure 10. Structural scheme of the system of the aircraft scattering fields mapping.

4.2. Comparative analysis of created imaging techniques

4.2.1. Spot visualization

Figure 11 (a) shows an example of point imaging of an electromagnetic field as the easiest implementation of this procedure. This technique is more productive, but less informative. The researcher does not see the values that appear between the two calculated points. This visualization is more difficult to understand, when it is superimposed on the image of the electronic device under study.

4.2.2. Cubic visualization

More advanced is the cubic imaging of the electromagnetic field shown in Figure 11 (b).

In this case, cubes are used instead of dots, the sides of which are equal to the distance between calculated points. The surface of the cube has transparency, depending on the magnitude of the measured field. With this approach, the researcher sees sharp transitions between the two calculated values, and artifacts may be observed under some viewing angles. At high resolution (number of points) of the calculated volume, the number of frames, processed per second, can be decreased, which in some cases may be critical. Changing the cubes position in space leads to incorrect visualization. This technique is less productive than the point one.

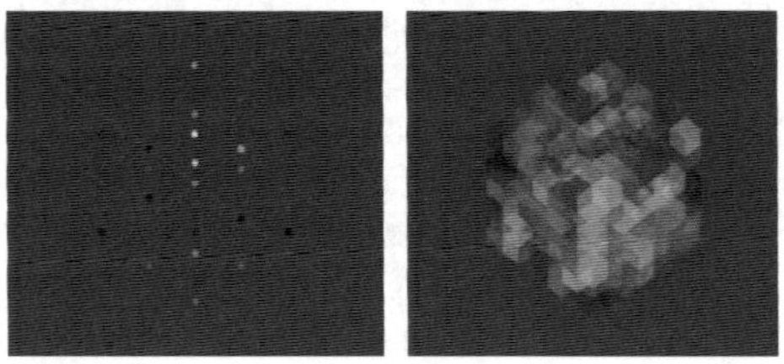

(a) (b) **Figure 11.** Point-like (a) and cubic (b) visualization.

4.2.3. Spherical visualization

This technique, illustrated by Figure 12 (a), is a cubic one extension, but instead of cubes, spheres are supposed to be used here. The sizes of the spheres are several times greater than the distance between the calculated points. This is necessary to ensure that the spheres are being overlapped. The surface of

IOP Conf. Series: Materials Science and Engineering 1027 (2021) 012027 doi:10.1088/1757-899X/1027/1/012027

the spheres also has transparency, depending on the field induction at the point within the sphere. This transparency allows to mix the colors of the spheres, forming more correct values at adjacent transitions, than in the case with cubes. The spherical imaging technique is less productive than the cubic one but lacks some of its drawbacks.

4.2.4. Flat visualization
Unlike previous methods, this technique does not link the geometric shape with each point of measurement. Instead, it is proposed to use a plane, on which the induction of the electromagnetic field is being measured, as shown in Figure 12 (b). A more accurate method, the bi-linear interpolation, is used to find the characteristic of the field between the nodes of the grid. This technique is devoid of artifacts. The number of frames, processed per second, does not depend on the calculated volume resolution. Any geometric 3D shape consisting of planes, can also be easily used in this technique. Besides, the technique supports the visualization of several planes at the same time, which allows to form a visualized volume as 3D visualization.

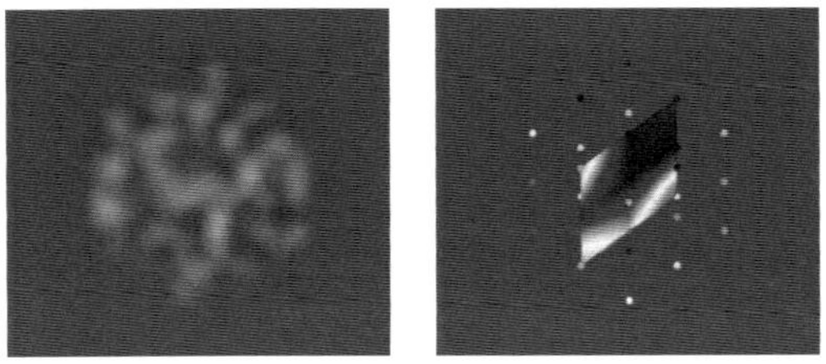

(a) (b) **Figure 12.** Spherical (a) and flat (b) visualization.

4.3. Benefits and prospects
The proposal provides visualization by using the brightness and color of markers. The software product offers the user a choice — visualization on a plane, or in 3D-space. The proposed technique allows to restore dependence by a minimum number of points, which reduces the number of measurements. An applications of the proposed system consist in different approaches to the analysis of the outer fields created by the aircraft. Moreover, the fields under study can be any, not only electromagnetic — temperature, fields of gravity, fluid flow rate, fluid speed and others. One of the promising applications is the mapping of the electromagnetic field in the cockpit of an advanced combat destroyer. Figure 13 shows the cockpit mapping grid used in fighter fire control technologies [24]. The figure shows that the cockpit of the aircraft is divided into cubic elements with a side of several centimeters, which determines the required accuracy of positioning of the measuring device, and the accuracy of visualization respectively.

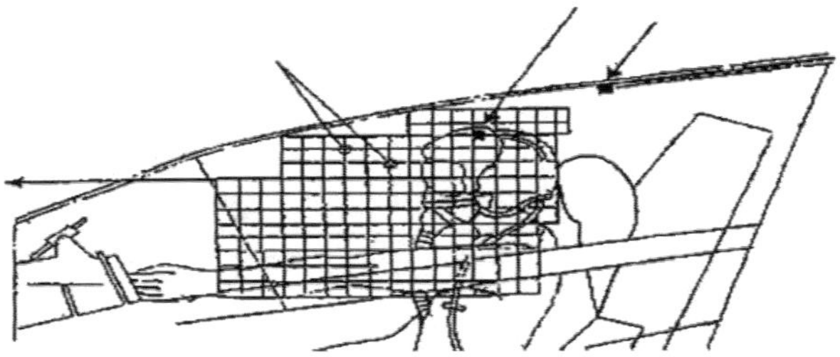

Figure 13. Dividing the cockpit work area into discrete cells.

5. Conclusion
This work implements an algorithm of flat and 3D visualization in the space and time of the electromagnetic scattering field induction created by aircraft in the surrounding space of a given volume. The formulas and rules have been obtained to determine both the position of the measurement results and their order of presentation in the general array. The visualization options are considered, the algorithm of bi-linear interpolation is implemented on the example of calculating the vector of

electrical field induction. A method of 3D display of measurable information is proposed. The algorithm of the initial definition of the number and position of points in the volume of the observation zone is realized. A color map is offered — a scale of measured values. It is possible to use a system of mapping scattering fields of power systems during research and development of aerospace technologies. In the next stages, the authors will develop the results in the direction of further automation of measurement processes, expanding the initial dialogue with the software product, and creating a permanent prototype system to demonstrate and improve it. In the next work, the authors plan to describe the results of mathematical and physical simulation of the proposed technology, for which the choice of object and preparation of equipment are carried out.

References

[1] https://portal.tpu.ru/SHARED/u/ULTRATONE/uchebnaya_rabota/Tab/Lecture.pdf
[2] http://venec.ulstu.ru/lib/disk/2017/102.pdf
[3] https://studme.org/179273/matematika_himiya_fizik/elementy_metoda_setok_primery_raznostnyh_shem
[4] https://wiki.org/Visualization
[5] https://skomplekt.com/harakteristiki-ostcillograf/
[6] Shnirman G N 2003 *Apparaturnye nabljudenija. Izbrannye trudy* (Moscow: OIFZ RAN)
[7] http://xn--90ahjlpcccjdm.xn--p1ai/produkcija/registratory/
[8] https://netology.ru/blog/7-services-data-visualization,
[9] https://freelance.today/poleznoe/20-luchshih-instrumentov-dlya-vizualizacii-dannyh.html,
[10] https://soware.ru/categories/data-visualization-systems
[11] Detlaf A A, Javorskij B M and Milkovskaja M B 1977 *Kurs fiziki* Vol 2 (Moscow: Vysshaja shkola)
[12] Kalantarov P L and Nejman L R 1951 *Teoreticheskie osnovy jelektrotehniki* (Gosjenergoizdat)
[13] Frish S and Timoreva A B 1962 *Kurs obshhej fiziki* (Moscow: GIFML)
[14] Korn G and Korn T 1973 *Spravochnikpo matematike* (Moscow: Nauka)
[15] https://habr.com/ru/post/342708/
[16] Samarskij A A and Gulin A W 1989 *Chislennye metody* (Moscow: Nauka)
[17] http://www.mtjournal.ru/archive/2017/meditsinskaya-tekhnika-3/sverkhchuvstvitelnyy-vektornyy-magnitometr-dlya-kartograficheskikh-izmereniy-v-kardiografii
[18] https://mipt.ru/education/chair/physics/S_III/lab_el/Shelagin.pdf
[19] https://www.advantech.ru/products/1-2mlkc9/pci-1716/mod_86ec4c4d-f497-45c5-81da-b8600c0eb36f
[20] Raab F H 1977 *Remote object position locater* Patent 4054881 USA G01S 3/02 The Austin Company
[21] Raab F H 1982 *Remote object position and orientation lacater* Patent 4314251 USA G01S 5/02 The Austin Company
[22] Egli W H et al *Helmet-mounted sighting system* Patent 4287809 USA F41G 9/00 Honeywell inc. 1981.
[23] Zimmer P et al 1995 *Method and apparatus for determining an orientation associated with a mobile system, especially a line of sight inside a helmet visor* Patent 5457641 USA G01B 7/04 Sextant Avionique

IOP Conf. Series: Materials Science and Engineering 1027 (2021) 012028 doi:10.1088/1757-899X/1027/1/012028

Synthetic pseudo-random sequence

A A Tsatsin

Moscow research and production complex "Avionika" named after O.V. Uspenskiy, 7 Obraztsova, Moscow 127055, Russia

sand_e@mail.ru

Abstract. The paper considers the basic information about a new class of pseudo-random sequences — the class of synthetic pseudo-random sequences. The main distinguishing feature of the sequences under consideration is the extended observation window. The proposed synthetic sequences are an extension of the area of existing pseudo-random sequences.

1. Introduction

Pseudo-random sequences (PRS) are widely used in modern communication systems, positioning systems, and instrumentation systems. Linear PRS: Gold sequences, Kasami sequences, M-sequences, and nonlinear PRS: de Bruijn sequences, GMW sequences and etc., are most widely used.

One application of pseudo-random sequences in instrumentation is the formation of digital encoder scales. The scales of binary single-track digital encoders, for example, based on the M-sequence [1], the de Bruijn sequence (T-sequence) [2], are based on one of the main properties of the PRS — the «window» of observation [3, 4]. The «window» property is that if a window from the number of neighboring characters, the width of which corresponds to the bit depth of the scale, is moved along the PSP period, then only unique combinations of numbers will be observed in this window [5].

This paper considers a new class of pseudo-random sequences with an extended observation window.

2. Basic information

To preserve the «window» property in full, i.e. to cover all possible unique combinations of codes, the maximum length of the numeric sequence N_S in the number system with base b for a given bit width m must be determined by the expression from combinatorics, as «placement with repetition»:

$$N_S(b) = b^m. \quad (1)$$

For parametric estimation of window introduces parameter extend the window of observation k_w — shift between adjacent assesses the elements of the observation window [6].

Existing classical pseudo-random sequences have a window of observation that evaluates neighboring quanta of the code scale (Figure 1). The extended observation window has the form (Figure 2).

One of the possible solutions to preserve the main "window" property is to synthesize special synthetic pseudo-random sequences with an extended observation window.

IOP Conf. Series: Materials Science and Engineering 1027 (2021) 012028 doi:10.1088/1757-899X/1027/1/012028

Figure 1. The classic observation window (k = 1).

Figure 2. The extended observation window (k = 2).

In [6], the concepts of aliquot and synthetic sequences are introduced and operations over them are described.

Aliquot sequences — a set of n cyclic sequences of length N_A with the observation window parameter $k_w = 1$, which correspond to n disjoint contours of the de Bruijn graph of order $N_S = N_A \cdot n$, containing all the vertices of the graph.

Synthetic sequence — a pseudo-random sequence of length N_S with an observation window parameter $k_w = n$, obtained by performing a merge operation on a set of n aliquot sequences of length $N_A = N_S / n$.

Figure 3 shows an example of a de Bruijn graph (G_B graph) of order $N_S = 8$, corresponding to a given bit width $m = 3$ for $b = 2$ [7].

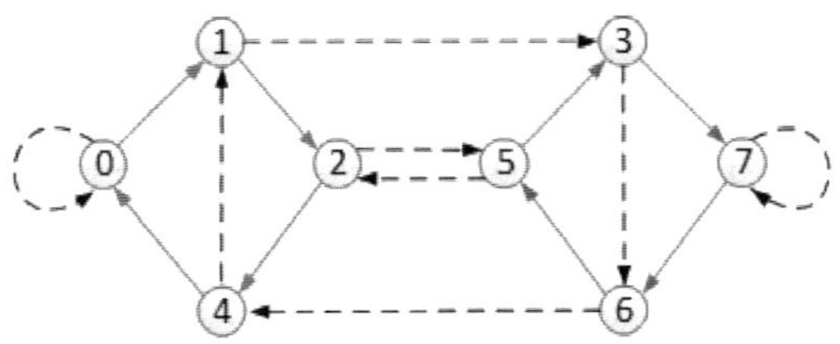

Figure 3. G_S — graph for b = 3, m = 3.

In the graph G_B, two disjoint contours are highlighted in red and blue. For contours starting from vertices with numbers 1 and 3, we apply the division operation modulo 2 to the vertex numbers of the contours. We get two aliquot cyclic binary A-sequences:

$$A_1^2 = \{1000\},\ A_2^2 = \{1101\}.$$

After performing the merge operation [6] on the existing aliquot sequences, we get a synthetic pseudo-random sequence:

$$S_2^2(3) = \Uparrow \left\{ A_1^2 A_2^2 \right\} = \left\{ 11010001 \right\}.$$

This synthetic pseudorandom sequence has the window extend parameter $k_w = 2$ (see Figure 2). If such an extended window is moved along the received PRS, then in accordance with expression (1), with the accepted parameter values $(m = 3, b = 2)$, we will observe eight unique codes. This confirms that the properties of the "window" are preserved when it is extended.

The example also shows that the parameter k_w has a dual role. It is the aliquot coefficient, i.e. how many of A-sequences a synthetic S-sequence consists of.

Note that in accordance with the definitions of multiple and synthetic sequences introduced, pseudo-random de Bruijn sequences [7] must be considered as a special case of synthetic pseudorandom sequences. This position follows from the extended interpretation of the theory of pseudorandom sequences considered in [6] and the corresponding extension of the theory of the conceptual apparatus.

Synthetic pseudorandom sequences longer than $N_S \geq 32$ characters are of more interest (table 1).

Table 1. Examples of binary *A*- and *S*-sequences, $N_S = 32$.

№	k_w	*A*-sequence	*S*-sequence
1	1	00000100011001010011101011 011111	00000100011001010011101011011 1 11[a]
2	2	0000011000101001 0101110011111011	0001000101111000010111011100 01 11
3	4	00001010 00011011 00100111 11111010	000100010011010111010010111101 10

[a] de Bruijn sequence.

Table 1 shows that for the same length, there are synthetic pseudo-random *S*-sequences that are based on sequences of different aliquot. In this case, all the given *S*-sequences are different, and they retain the "extended window" property.

Further research has shown that synthetic pseudo-random sequences can be obtained for other number systems (ternary, quaternary, etc.).

The most complete conceptual apparatus of open synthetic pseudorandom sequences is described in [6].

3. Summary

Thus, the article considers a new class of pseudorandom sequences — synthetic pseudorandom sequences. The main distinguishing feature of the sequences under consideration is the extended observation window. The discovery of sequences of this class significantly complements the theory of pseudorandom sequences, and allows us to expand scientific research in this area.

References

[1] Ozhiganov A A, Pribytkin P A, Pavlov V V et al Patent RF № 2510572 29.10.2012 Psevdosluchaynaya kodovaya shkala 27.03.2014 No 9

[2] Tsatsin A A 2017 Metod formirovaniya T-posledovatel'nostey dlya odnokoordinatnykh tsifrovykh datchikov peremeshcheniy *Materialy IV nauchno-prakticheskoy konferentsii pamyati O V Uspenskogo* (Moscow: Izdatel'skiy dom Akademii N Ye Zhukovskogo) p 54

[3] Fredricsson S A 1975 Pseudo-Randomness Properties of Binary Shift Register Sequences *IEEE Trans. Inform. Theory* **21** 115–20

[4] Blackburn S A Note on Sequences with the Shift and Add Property 1996 *Designs, Codes, and Crypt.* **9** 251–6

[5] MacWilliams J F and Slone Neil J 1976 A Pseudo-Random Sequences and Arrays *Proc. of The IEEE* **64** 12

[6] Tsatsin A A Sinteticheskiye psevdosluchaynyye posledovatel'nosti dlya datchikov polozheniya 2019 *Pribory* **11** (233) 21–8

[7] de Bruijn N G 1946 *A combinatorial problem* (Koninklijke Nederlandse Akademie van Wetenschappen) **49** 758–64

IOP Conf. Series: Materials Science and Engineering 1027 (2021) 012029 doi:10.1088/1757-899X/1027/1/012029

Improve the integrity testing process based on QFD, FMEA and TRIZ

M V Vysotskaya and A Ya Dmitriev

Samara National Research University named after academician S.P. Korolev (Samara University), 34 Moskovskoye Shosse, Samara 443086, Russia

Manya_93@mail.ru

Abstract. In today's aircraft industry — competitiveness in the domestic and foreign markets, is the main criterion for the development of quality, a high level of which is achieved through methods and approaches: QFD, FMEA, TRIZ. There is a possibility of application of these methods at tests of production, for the purpose to meet expectations of consumers with the most optimum combination of characteristics of production. The design process is often associated with a compromise of conflicting parameters, which means increased costs and loss of quality. Innovation in the production process can create great value. The rapid development of modern technologies and the commercial environment suggests that the design of production processes requires the application of a systematic method of improvement. The proposed method combines a number of quality tools to support the process of modern development of aviation. In order to implement the acceptance improvement, it is necessary, first, to analyze the product testing process in accordance with the influencing parameters and use QFD (quality function deployment) to obtain the basic requirements for the testing process. Secondly, to find the contradictions of the testing process and analyze them comprehensively. To creatively solve the revealed contradictions, apply the Russian theory of inventive problem solving (TRIZ). When using this technique increases both the efficiency and maneuverability of the use of quality tools to solve technological problems. The effectiveness of the proposed method of improvement is illustrated and confirmed by the example of the process of testing the tightness of the hose of the life support system of the product "Bion – M".

1. Introduction

Aircraft belong to complex technical systems, failures of which often lead to serious consequences.

Production of any production is based on stability of technological processes which confirmation serve static and dynamic tests of production with the maximum approach to real operating conditions. Static tests to a greater extent characterize the properties of the material, and dynamic — the validity of the selected parameters of structures and technologies.

Operational testing plays an important role in the production life cycle, as it provides an opportunity to confirm compliance with the requirements that should be assessed in conditions as close as possible to the actual operation.

The main methods of conformity assessment are measurement, testing and control, which must be carried out in accredited laboratories with certified testing equipment.

It is advisable to improve the quality of products to use modern tools and methods of quality management, such as: QFD, FMEA, TRIZ.

FMEA — “Analysis of the types and consequences of potential nonconformities” assesses the risks and possible damage caused by design and process nonconformities at an early stage in the design and construction of the finished product or components. Is divided into: PFMEA to identify potential inconsistencies of the process and DFMEA to identify potential inconsistencies of the design [4].

QFD — “quality function deployment Method”, provides high quality of the created products at each stage of the life cycle with a guarantee of obtaining the final result that meets the requirements and expectations of the consumer [5].

TRIZ — "Theory of inventive problem solving", used to identify and use the laws, laws and trends in the development of technical systems. TRIZ is designed to organize the creative potential of the individual so as to promote self-development and the search for solutions to creative problems in various fields. The main task of TRIZ is to propose an algorithm that allows to find the most suitable variant without searching for infinite variants of solutions to the problem, discarding the less qualitative ones. Or, to put it more simply, TRIZ allows you to solve the inventive problem so that the output to get the highest efficiency. [1]

According to modern research in the aerospace industry, the test process plays an important role between design development and production. However, the testing process has limitations or contradictions and requires a compromise between many factors. The rapidly changing modern market makes us strive for high competitiveness and develop production processes with the use of systematic improvement techniques to ensure stable product quality. A number of quality tools, methods and approaches are used to improve test design processes. Innovation processes usually mean a change in test technology that includes new designs, processes, the use of modern equipment and test models. This article describes the design strategy of production processes on the example of combining modern quality management tools: QFD, FMEA and TRIZ [2].

QFD and FMEA can help to identify the requirements to test the system and determine actions to address the shortcomings of the system. The TRIZ method is used to improve the testing process and to solve the conflict parameters of the process system. Since the design process at an early stage plays a crucial role, which determines the degree of compliance of the product with the declared requirements of the consumer, it is extremely important to build a systematic approach to the design of the test process. The implementation of a compromise strategy in the design of test processes helps to recognize and manage conflicting process parameters and find creative solutions to the problem [6].

TRIZ is a theory that can effectively resolve technological conflicts or contradictions with the help of original ideas. In the design of the test process and the use of TRIZ analysis of the contradictions of the process, which are the source of conflict moments of the test environment, is inevitable. However, if you solve the problem comprehensively, especially in a complex system of production processes, it is necessary to learn about all the requirements, not only about the demands of consumers, but also the requirements for the test system. And QFD (quality function deployment) can be an effective tool for requirements analysis.

In accordance with the above, this article proposes a strategy for the improvement process. QFD, FMEA and TRIZ are used as an integrated methodology to select the appropriate solution for emerging problems in the testing process. First, QFD is used to transform process requirements into technical specifications. Meanwhile, many technical contradictions can be found, given the complexity of the system of production processes. In this case, TRIZ is used, for example, standard methods of eliminating technical contradictions, which are one of the effective tools to break down the complex system of the design process into principles and achieve several alternative modern solutions. Common design parameters for selecting a suitable process system include productivity, safety, environment, quality, flexibility and cost. If you cannot achieve a solution to the problems, you must apply TRIZ again. An example of the improvement process strategy is shown in figure 1 [3].

A methodology for the systematic and structured transformation of customer requirements into product requirements and the further deployment of requirements down to the workplace with the least risk is a modern instrumental way of developing innovations.

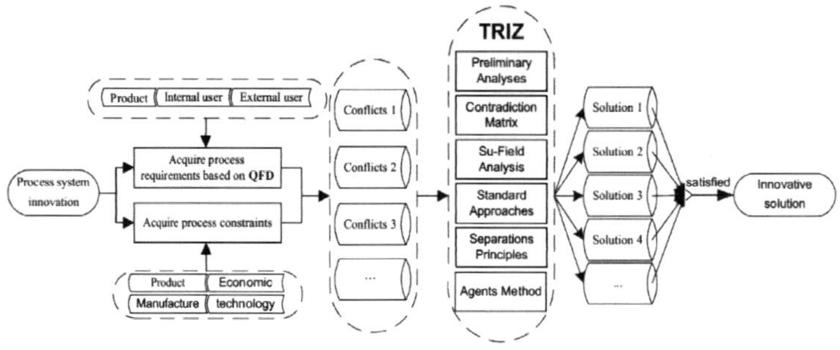

Figure 1. Improvement process strategy.

2. Object of research

The strategy of the improvement process is applied on the example of testing the tightness of the hose of the life support system of the product 12XM. Tests are carried out on the pneumatic panel VK 70209-398A. The block diagram of the test bench is shown in the figure 2.

The object of the test is the hose of the life support system of the product 12 KSM presented in figure 3. Hose construction consists of seal (pin & sleeve) and the sleeves 2K7-30, supplied by THAT 38.0051515-92. The elements of the sealing hose is exposed to the greatest stress, as it affects them the most free weight of the sleeve 2K7-30. The hose is installed on the outer surface of the spacecraft on the cable mast and serves to supply oxygen into the spacecraft from oxygen cylinders.

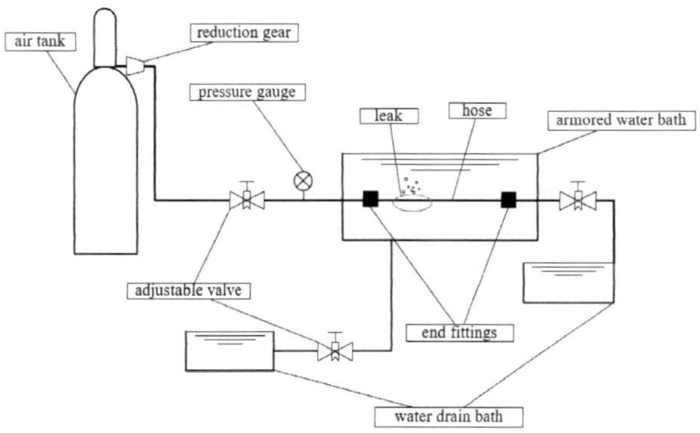

Figure 2. Structural diagram of the test bench.

Based on the test procedure program, there are test requirements.

1. Before each test:
   - inspect hoses for mechanical damage and contamination;
   - in the case of the presence of moisture on the hoses to keep them dry.
2. After each test, the hoses must be completely drained and kept plugged. Drying is carried out by blowing air until complete removal of moisture.
3. When installing the hoses in a device for testing the tightness of coupling nuts of the joints to produce 1.0 to 1.5 facets of focus, are not allowed twisting of the hoses around the axis. Seal the joints with gaskets.
4. Test the hose for tightness with compressed air pressure (3.45 ± 0.15) MPa [(35 ± 1.5) kgf/cm^2] for 5 minutes under water with a preliminary exposure of 10 min. No tightness is not allowed (no air bubbles from the hose in the water and pressure drop on the pressure gauge).
5. The necessary measuring instruments used in the tests must be verified and the test equipment must be certified.

3. Application of the QFD method to identify the basic requirements and process parameters «leak test»

The term QFD is a method that allows to deploy the customer requirements into performance characteristics, performance characteristics — into part characteristics, part characteristics into manufacturing process and manufacturing process — into control tools [6, 7]. QFD should be used since the conceptual design stage, because the decisions made at this stage provides about 70% of the project's success with minimal costs [8].

Figure 4 shows the structure of the main QFD tool — the House of Quality, which received such a name for its shape.

Requirements for the process of testing for tightness and contradictions of parameters exist simultaneously, which usually affects the appearance of many technological conflicts between the accuracy of observations and the speed of testing, the presence of air bubbles and the lack of pressure, etc. it is Necessary to assess the requirements and contradictions of the test process, this requires to find out the controversial points of the process without missing anything. The first thing to do is to find out the exact parameters of the requirements, not a General description. QFD (quality function deployment) can be an effective tool for this. Secondly, the limitations of the process system, which are related to costs, technologies and resources, etc., must be fully taken into account.

The purpose of the deployment of quality functions (QualityFunctionDeployment — QFD) is to ensure the quality of the created products at each stage of the life cycle, which would guarantee the final result that meets the requirements and expectations of the consumer [7].

Technology QFD includes a 4-phase (level): quality plan product (phase 1), the quality of the product components (phase 2), quality plan process (phase 3), the quality plan of operations (phase 4), each of which used a matrix diagram is a special kind of quality. Quality house is an element of quality functions deployment technology (QualityFunctionDeployment — QFD) [8].

To identify the quality requirements of the process and operations, we need the third and fourth level of QFD methodology. For QFD the third level, we need requirements to test the system and steps of the process of "testing for leaks".

Requirements for the test system according to the degree of failure are divided into three main types:

1. **Anticipatory:**
    - ergonomics of testing.
2. **To health:**
    - the presence of the user in the workplace;
    - high performance (maximum number of tests per day);
    - the safety of good products.
3. **Obligatory:**
    - quality assurance seal of rubber hoses to end fittings;
    - impact of loads provided by operational factors;
    - reproducible taking into account the capabilities of the test equipment;
    - evaluation of the reliability of the test object according to the test results;
    - security testing.

The stages of the process "test for leaks»:

1. Install the hoses in the equipment and connect to the pneumatic console;
   To drown out the hoses;
2. Give pressure to hoses with compressed air P = 0.5 kgf/cm^2;
   Place the hoses in an armored bath water;
3. To increase the pressure in the hoses to 36 kgf/cm^2;
   Withstand pressure for at least 10 minutes;

4. Check for the tightness of the sealing space and the wall of the hose within 5 minutes. Leakage is not allowed (Signs of leakage are air bubbles in the chromatic solution or pressure drop on the pressure gauge);
5. To relieve the pressure to 0.5 kgf/cm^2;
6. Remove hoses from armored bath;
   Bleed the pressure to 0 kg/cm^2.
7. Disconnect the hoses from the pneumatic console;
   Remove plugs from hoses and tape;
8. Blow the hose with compressed air P = 10 kgf/cm^2 until the moisture is completely removed;
9. To reflect the result of the test;
10. To establish technological plugs in the hose.

Figure 4 shows the third level QFD.

As a result of the QFD of the third level, the key operations and parameters of the "leakproofness test" process are determined»:

1) Withstand pressure for at least 10 minutes (14.5%);
2) Reflect the result of the test in the sheet of test marks (11.5%);
3) Install the hoses in the equipment and connect to the air gun (10.6%).

For QFD the fourth level, we need the operation to the process "test hose for leaks".

Auxiliary requirements for the "leak test" for the hose object of the life support system of the product 12 KSM are presented in the second section "Object of study". Figure 5 shows the QFD of the fourth level.

Based on the matrix diagram, QFD allows you to understand the key operations that affect product performance, product quality and process parameters of the "leak test".

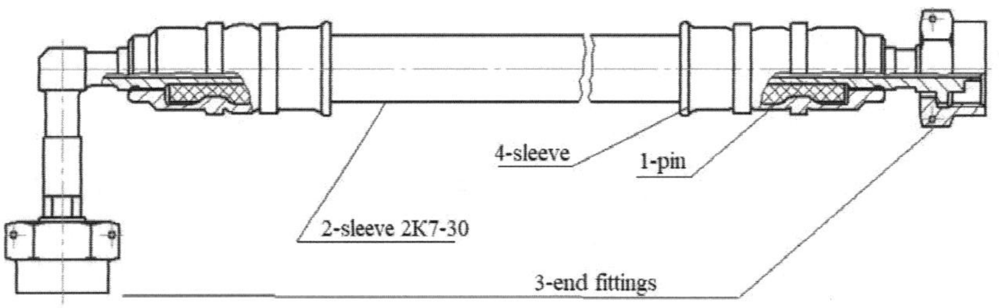

Figure 3. Hose, life support system.

Row #	Max Relationship Value in Row	Relative Weight	Weight / Importance	Quality Characteristics (a.k.a. "Functional Requirements" or "Hows") / Demanded Quality (a.k.a. "Customer Requirements" or "Whats")	Install the hoses in the equipment and connect to the pneumatic panel	Plug the hoses	Give the pressure in the hoses with compressed air 0.5 kgf / cm2	Place the hoses in an armored water bath.	Increase the pressure in the hoses to 36 kgf / cm2	Hold under pressure for at least 10 minutes	Check tightness of the place of seal and wall - 5 minutes	Relieve pressure to 0.5 kgf / cm2	Remove hoses from armored bath	Relieve pressure to 0 kgf / cm2	Disconnect the hoses from the pneumatic panel	Remove plugs and cassette from hoses	Blow out the hose with compressed air 10 kgf / cm2 to remove moisture	Reflect the test result	Install process plugs on the hose
1	9	18,5	10,0	Checking the quality of sealing rubber hoses in the end fittings under the influence of loads	O					⊙									
2	9	18,5	10,0	Evaluation of the reliability of the test object according to test results		O			⊙	⊙	⊙			⊙			⊙	⊙	
3	9	18,5	10,0	Safety test	⊙										⊙	⊙			⊙
4	9	14,8	8,0	Safety of tested tests	▲	▲	⊙			⊙	O	⊙		⊙			⊙	⊙	
5	9	13,0	7,0	The presence of instructions in the workplace	⊙	▲	⊙			O	O	⊙							▲
6	9	11,1	6,0	High performance (maximum number of tests per day)				O										⊙	
7	9	5,6	3,0	Ergonomic testing	O								O		⊙				▲
				Target or Limit Value			0.5 kgf / cm2	armored bath with water	up to 36 kgf / cm2	at least 10 minutes	5 minutes	up to 0.5 kgf / cm2		0 kgf / cm2			purge with a pressure of 10 kgf / cm2	test mark sheet	
				Difficulty (0=Easy to Accomplish, 10=Extremely Difficult)															
				Max Relationship Value in Column	9	3	9	3	9	9	9	9	3	9	9	9	9	9	9
				Weight / Importance	370,4	83,3	250,0	33,3	166,7	505,6	250,0	250,0	16,7	300,0	216,7	166,7	300,0	400,0	185,2
				Relative Weight	10,6	2,4	7,2	1,0	4,8	14,5	7,2	7,2	0,5	8,6	6,2	4,8	8,6	11,4	5,3

Figure 4. Third level QFD for «leak test».

IOP Conf. Series: Materials Science and Engineering 1027 (2021) 012029 doi:10.1088/1757-899X/1027/1/012029

Row #	Max Relationship Value in Row	Relative Weight	Weight / Importance	Quality Characteristics (a.k.a. "Hows") / Demanded Quality (a.k.a. "Whats")	Inspect hoses for mechanical damage and impurities	If there is moisture on the hoses, make them dry	After each type of test, the hoses must be completely drained	Hoses should be kept completely plugged	Air drying is carried out to completely remove moisture	To make a cape of nuts on 1,0-1,5 sides from an emphasis	Test with compressed air pressure of 3.45 ± 0.15 MPa (35 ± 1.5 kgf / cm2)	Test for 5 minutes under water at 10 minutes pre-exposure	Means of measurement are verified, test equipment certified
1	9	10,6	370,4	Install the hoses in the equipment and connect to the pneumatic panel	O	▲		▲			O		Θ
2	9	2,4	83,3	Plug the hoses				Θ		O			
3	3	7,2	250,0	Give the pressure in the hoses with compressed air 0.5 kgf / cm2	O						O		
4	9	1,0	33,3	Place the hoses in an armored water bath.	Θ								
5	9	4,8	166,7	Increase the pressure in the hoses to 36 kgf / cm2							Θ		
6	9	14,5	505,6	Hold under pressure for at least 10 minutes								Θ	
7	9	7,2	250,0	Check tightness of the place of seal and wall - 5 minutes	▲							Θ	
8	9	7,2	250,0	Relieve pressure to 0.5 kgf / cm2							Θ		
9		0,5	16,7	Remove hoses from armored bath									
10	3	8,6	300,0	Relieve pressure to 0 kgf / cm2			O		O		▲		
11	1	6,2	216,7	Disconnect the hoses from the pneumatic panel									▲
12		4,8	166,7	Remove plugs and cassette from hoses									
13	9	8,6	300,0	Blow out the hose with compressed air 10 kgf / cm2 to remove moisture		▲	Θ		Θ				
14	9	11,4	400,0	Reflect the test result					Θ		O	O	Θ
15		5,3	185,2	Install process plugs on the hose									
				Target or Limit Value									
				Difficulty (0=Easy to Accomplish, 10=Extremely Difficult)									
				Max Relationship Value in Column	9	1	9	9	9	3	9	9	9
				Weight / Importance	69,0	19,2	103,0	32,1	206,0	7,2	203,5	228,9	204,6
				Relative Weight	6,4	1,8	9,6	3,0	19,2	0,7	19,0	21,3	19,1

Figure 5. QFD fourth level for the process of verification of impermeability of the hose.

4. The application of TRIZ

There are many controversial points in the testing process, so it is advisable to use TRIZ at an early stage of design. One of the prerequisites of TRIZ is that there are objective laws of development and functioning of systems, based on which it is possible to build inventive solutions. In other words, many technical, production, economic and social systems are developing according to the same rules and principles. Soviet scientist and inventor Heinrich Saulovich Altshuller discovered them by studying the patent Fund and analyzing the ways of development and improvement of technology for a long time. A list of 40 standard methods of eliminating technical contradictions has been created — a kind of desktop guide for the inventor, but a special kind of guide: the inventor should consider it as the basis for which it is necessary to replenish the new technical and patent publications. TRIZ is a human-oriented systematic methodology for solving inventive tasks based on knowledge. It can solve technical problems and offer innovative solutions using a knowledge base.

To study the similar features of other inventions for the use of TRIZ and determine the patent purity of the studied object of technology: a stand for checking products for leaks, a patent search was conducted and materials were selected for further analysis:

1) Stand tightness control of filter elements and filter packages — utility model №74617.
2) Stand for testing the tightness of the transformer housing — utility model №106368.
3) Stand for checking the tightness of radiators before and after repair — utility model №172904.
4) Stand for checking the tightness of gas valves of the boiler — utility model №99849.
5) Walls for leak test — the invention № 2583880.

Stand for testing air hoses — utility model № 38 463 is selected as an analogue for the use of TRIZ, the stand refers to devices for monitoring and testing of brake systems and is designed for testing brake hoses with air. Currently, the company is known to stand the test of hoses, which hoses the consolidation is done manually, the lifting and lowering of the frames with handheld hoses.

The disadvantage of the known stand is the use of manual labor, that is, manual hose clamping and lowering of the framework.

The purpose of the proposed utility model is to increase the performance of testing hoses by automating the testing process by using a pneumatic actuator and a sealing unit.

In this article TRIZ is applied in relation to the process of testing the hose of the life support system of the product 12KSM, the purpose of which is to check the quality of sealing of rubber hoses in the end fittings by the loads provided by the operational factors reproduced taking into account the capabilities of the test equipment, as well as to assess the reliability of the test object according to the test results.

Risk analysis by FMEA [4] allowed to determine a number of contradictions of the test process, for example, such as: the dependence of the pressure in the test bench, affecting the test result, i.e. if the pressure in the system is up to 36 kgf/cm^2, it is possible defective hose is allowed to work, at a pressure above 36 kgf/cm^2 there is the appearance of microcracks in the hose.

On the basis of the table of application methods of solving technical contradictions the selected factors: No. 11 Pressure and No. 27 Reliability. To solve these contradictions, standard methods of their elimination are proposed, such as: № 19 "Principle of periodic action", which consists in the transition from continuous to periodic (impulse) and № 35 "Change of physical and chemical parameters of the object", which allows to change the concentration or consistency.

On the basis of the above methods, measures have been developed to improve the stand for tightness control:

- Air supply will be carried out according to a certain law (pulse mode) short-term supply to the system is not constant, but variable, with a certain frequency and amplitude of oscillations.
- Adding a chemical reagent to the test environment of the bath to change the physical and chemical properties of the water, in order to improve the visualization of the detection of non-tightness of the hoses.

5. Conclusions and discussion

1. Operational testing plays an important role in the production life cycle, as it provides an opportunity to confirm compliance with the requirements that should be assessed in conditions as close as possible to the actual operation.

2. Modern methods and approaches are applied and investigated: QFD, FMEA, TRIZ, for improvement of process of tests on tightness.

3. QFD fourth level for the process of verification of impermeability of the hose allowed us to identify key activities that could affect efficiency of production.

4. Risk analysis by FMEA revealed a number of contradictions in the testing process, for which TRIZ was used.

5. A review of patents for existing analogues of the test bench to test products for leaks, as well as the application of the theory of inventive problem solving allowed to come to solutions of technical contradictions.

6. Measures to improve the test stand are defined.

References

[1] Duflou J R and D'hondt J 2011 Applying TRIZ for systematic manufacturing process innovation: the single point incremental forming case *Procedia Engineering* **9** 528–37

[2] Li T S and Huang H H 2009 Applying TRIZ and Fuzzy AHP to develop innovative design for automated manufacturing systems *Expert systems with applications* **36** (4) 8302–12

[3] Wang C, Zhao W, Wang J, Li X, Zhang K, Guo X and Zeng J 2015 The Optimizing Strategy of Systematic Process Innovation based on QFD TRIZ and AHP *Applied Mathematics & Information Sciences* **3** 1593–604

[4] Vysotskaya M V, Khryashcheva D V, Orlova S I, Tsyman V N, Russkikh I V and Shurshev M A 2018 Improving the process of leak testing with modern quality management methods *Quality and life* **3** (19) 16–24

[5] Dmitriev A Ya and Mitroshkina T A 2019 Improving the efficiency of aviation products design

based on international standards and robust approaches *IOP Conference Series: Materials Science and Engineering* **476** 012009

[6] Dmitriev A Ya, Vashukov Yu A and Mitroshkina T A 2016 *Robust design and technological preparation of the manufacturing of aviation equipment* (Samara: Samara University) p 76 [In Rus]

[7] Dmitriev A Ya and Mitroshkina T A 2016 The ontological model and the hybrid expert system for products and processes quality identification involving the approach based on system analysis and quality function deployment *ITM Web of Conferences* **6** 02005

[8] Dmitriev A, Mitroshkina T and Rogachev G 2017 Structural and parametric analysis of robust design quality of complex technical systems *ITM Web of Conferences* **10** 01001

Investigation of changes in the operator's state by analyzing the characteristics of blinking

V N Yurko[1], A B Uali[2] and A S Naukenova[2]

[1] State Research Institute of Aviation Systems, 7 Viktorenko street, Moscow 125319, Russia
[2] M Auezov South Kazakhstan University, Shymkent, Kazakhstan

marmotto@rambler.ru, wlad_T90@mail.ru

Abstract. This paper discusses in detail an approach using the method of automatic blink recognition for assessing the state of the operator using deep convolutional neural networks. It also analyzes the characteristics of blinking to detect the facts of loss of concentration.

1. Introduction
Analysis of publications shows that the problem of assessing the state of the operator based on the characteristics of the blinking of operators is currently relevant and promising. Training of operators on modern simulators is accompanied by an increase in the quality of professional activity and optimization of neuro-emotional reactions [1]. When assessing the level of training of operators, a systematic approach is important [2]. It is not enough to assess the operator's training level only by the success of the flight task. High performance can be achieved due to excessive muscle tension, functioning at the limit of mental and psycho physiological capabilities in the absence of developed and stable skills. At the same time, the reliability of professional activity can be quite low. This manifests itself when the conditions for completing tasks become more complex. The reliability of the operator's activity increases significantly when the results of the task execution become stable, and the psycho physiological reactions correspond to the complexity of the task being performed. Fast and correct reactions, good coordination of accurate and proportionate actions of the operator characterize the professional ability, which, along with other qualities, determines its reliability.

2. Results of the experiments

2.1. Eye blink detection
To conduct research to assess the functional state of the operator, special software was developed based on a convolutional neural network for recognizing the operator's blinks with subsequent analysis of the blink parameters [3, 4]. When planning the experiment, a video camera with a video recording frequency of 120 Hz was selected, filming was carried out in the visible range. System testing has shown that the software is capable of detecting the presence of blinking with high accuracy, recognizing the state of the eyes — eyes are open and eyes are closed (figure 1). The result of the work of the developed software is the recording of the state of the operator's eyes in a file synchronized with flight data for further in-depth analysis.

IOP Conf. Series: Materials Science and Engineering 1027 (2021) 012030 doi:10.1088/1757-899X/1027/1/012030

Figure 1. Results of neural network algorithm detecting the blinks of an operator. The open eyes state; the closed eyes state.

2.2. Analysis of the blink characteristics

For operators who have normal run, the blink rate was 14–18 blinks per minute, the duration of a blink is on average 160 milliseconds (figure 2).

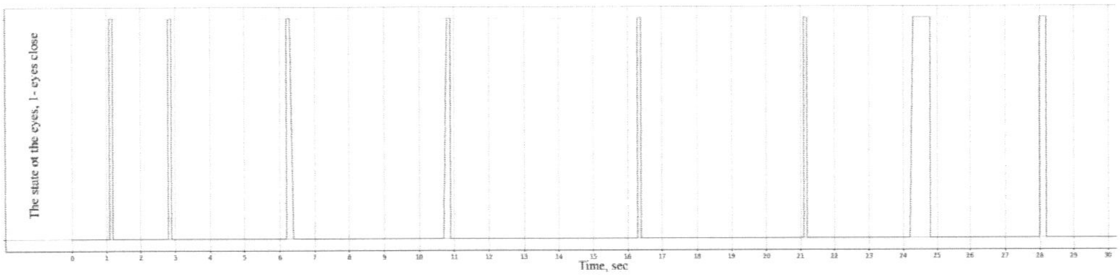

Figure 2. Output waveform of blinks during quiet work.

During the hard work on the flight simulator (the *manual* landings), the frequency of blinks decreased to 5–8 blinks per minute. Blinks became faster (60 milliseconds). With increasing consentration, the number of incomplete blinks also increases. This is due to the need for continuous monitoring of the flight parameters displayed on the simulator indicators (figure 3).

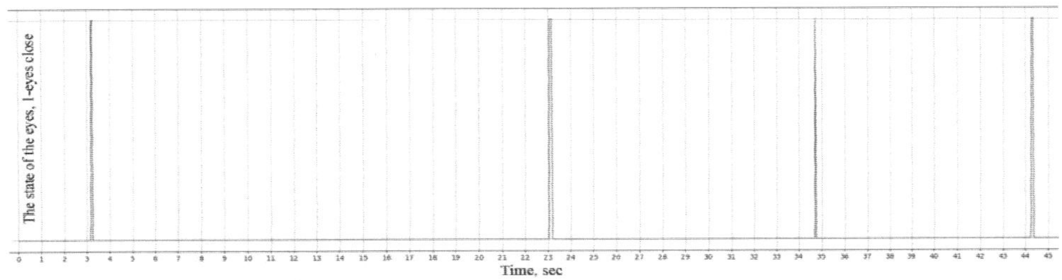

Figure 3. Output waveform of blinks when operator lands.

The experiments revealed correlations between blink parameters and piloting accuracy characteristics. Figure 4 shows the number of blinks per one landing against standard deviation of the vertical coordinate. With better piloting accuracy, a greater number of operator blinks per landing is

IOP Conf. Series: Materials Science and Engineering 1027 (2021) 012030 doi:10.1088/1757-899X/1027/1/012030

recorded, which corresponds to a normal run. The increase of errors means that the situation is getting out of control, which requires more efforts to restore the normality. In these experiments the duration of a landing was usually 3 minutes.

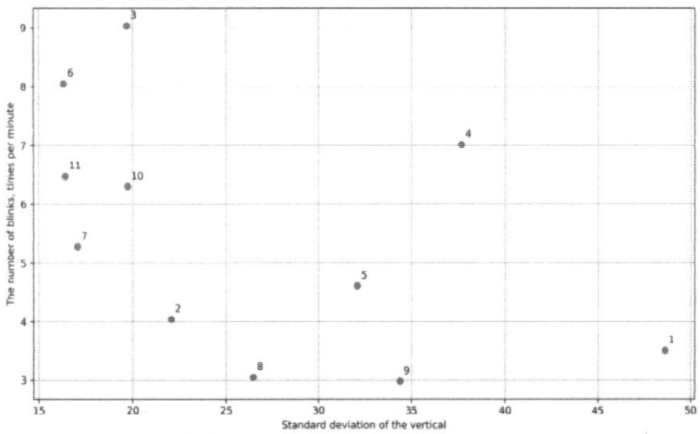

Figure 4. Plot of the number of blinks per landing against standard deviation of the vertical coordinate

Dependence on the number of blinks with an increase in the landing approach number, that is, with an increase in the time of continuous intensive work of the operator, needs to be investigated further. The task is complicated by the fact that two opposite tendencies are at work. On the one hand, the frequency of the operator's blinking decreases due to the perception of new information or strong concentration of attention, on the other hand, the frequency increases with the accumulation of operator fatigue. In figure 5 the decreasing trend prevails but there are exclusions in approaches 13 and 15.

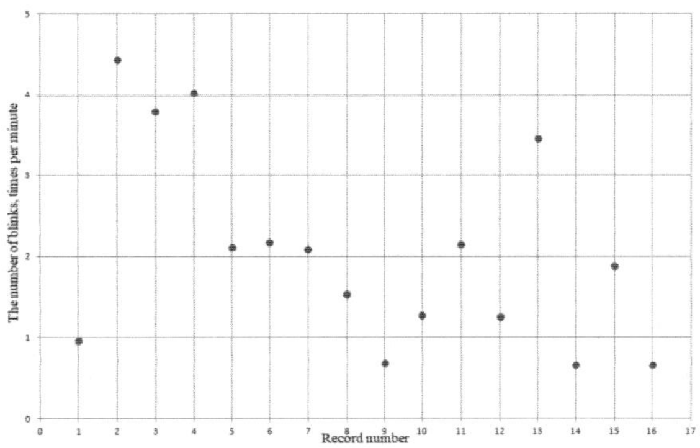

Figure 5. Graphs of the dependence of the number of blinks on the approach number. Eye and mucous membrane small irritation may result in an increase of the blinks number (figure 6).

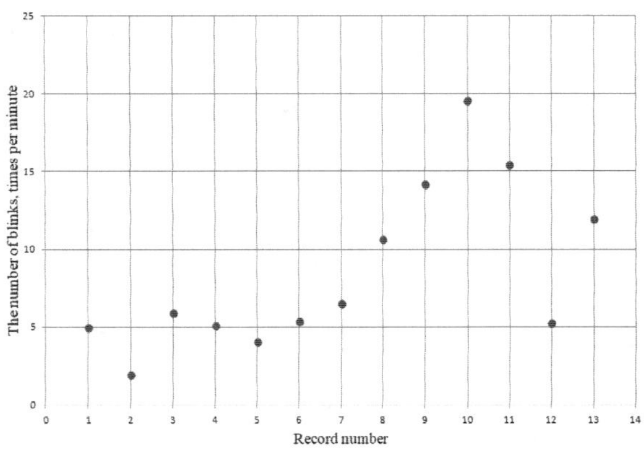

Figure 6. Graphs of the dependence of the number of blinks on the approach number in case of eye irritation.

The complexity of the problem of assessing the state of the operator is due to the fact that objectively measured parameters, for example, the characteristics of blinking, depend on the operator's properties in terms of perception of the surrounding situation, reaction, attentiveness, situational thinking and other psycho physiological characteristics and capabilities of the operator.

3. Conclusion

During intense work of operators, the average frequency and duration of blinks is greatly reduced — from 14–18 blinks per minute with an average duration of 160 ms during normal operation to 5–8 blinks per minute with an average duration of 60 ms in an intense mode. With increasing voltage, the number of incomplete blinks also increases. Long-term piloting in such modes contributes to severe fatigue. Also, a relationship was found between the accuracy of piloting and the blinking parameters. With a higher piloting accuracy during a landing approach a greater number of blinks is observed.

Acknowledgements

This work was supported by Russian Foundation for Basic Research (RFBR), project 18-08-01142-a.

References

[1] Yurko V N, Tsvetkov D V and Golovkin Y A Development of software — hardware simulators for airborne system for the flight simulators 2020 *Issues of Avionics* **2** (48) 27–39

[2] Korsun O N, Yurko V N and Mikhaylov E I 2020 *IOP Conf. Ser.: Mater. Sci. Eng.* **714** 012012

[3] Yiu Y H, Aboulatta M, Raiser T, Ophey L, Flanagin V L, zu Eulenburg P and Ahmadi S A 2019 DeepVOG: Open-source pupil segmentation and gaze estimation in neuroscience using deep learning *Journal of Neuroscience Methods* **324** 108307

[4] Ronneberger O, Fischer P and Brox T 2015 U-net: Convolutional Networks for Biomedical Image Segmentation *CoRR* abs/1505.04597

Improving the quality of FDM 3D printing of UAV and aircraft parts and assemblies by parametric software changes

R S Zagidullin, N I Zezin and N V Rodionov

Department of Aircraft Manufacturing and Quality Management in Mechanical Engineering, Samara National Research University, 34 Moskovskoye shosse, Samara 443086, Russia

zagidullin_radmir@mail.ru

Abstract. The article analyses and presents the classification of defects in parts obtained by FDM 3D printing technology. A fault tree has been built for the functional and structural analysis of the factors causing defects in parts produced by FDM 3D printing technology. Using the constructed fault tree, a method is proposed for eliminating defects in parts obtained by FDM 3D printing technology, based on parametric changes in software.

1. Introduction

Additive technologies (AT) are generating a paradigm shift, replacing traditional “subtractive” production technologies with “adding” ones. Such well-known manufacturing companies as Boeing, Airbus, SpaceX, Blue Origin, Rocket Lab and others are increasing the number of 3D-printed parts in their product designs every year [1]. AT allow to reduce production costs and costs at the stage of technological preparation of production is to exclude the development of a technological process and tooling, and, therefore, will reduce the cost of products and reduce the time to market.

Fused deposition modeling (FDM) is currently the most widespread and inexpensive type of AT. FDM 3D printing technology consists in building a 3D model from a molten polymer filament, which flows through the print head onto the desktop. FDM 3D printing makes it possible to obtain parts of almost any geometric shape from polymer materials. With the advent of composite and high-temperature polymer materials for 3D printing, FDM technology is widely used in aircraft construction, especially in the production of parts and assemblies for aircraft and unmanned aerial vehicles (UAVs) [2–4].

However, with FDM 3D printing, users of additive equipment, especially beginners, are faced with various defects that are caused by unsatisfactory chosen parameters of the 3D printing process, imperfections in the design of the 3D printer, and the preparation of the 3D model for printing. This paper proposes a method for improving the quality of FDM 3D printing by parametric changes in the software (SW) of additive equipment.

2. Application of FDM 3D printing in the production of UAVs and aircraft

With the advent of new polymer materials reinforced with carbon fiber and metal powders, FDM 3D printing technology is increasingly being used every year in the production of parts and assemblies for UAVs and aircraft. The advantage of FDM 3D printing in the production of UAVs and aircraft is the ability to use a bionic design method. Below are examples of FDM 3D printing applications in aircraft and drone construction.

Currently, in the aircraft industry, FDM 3D printing has found wide application for the manufacture of air ducts for aircraft and helicopters. Air duct assemblies and parts obtained using FDM 3D printers are used in Boeing 787 airliners, F / A-18, F-35 fighters and Bell 429 helicopters [2]. Figure 1 shows the details and units of the air ducts of the indicated aircraft and helicopters.

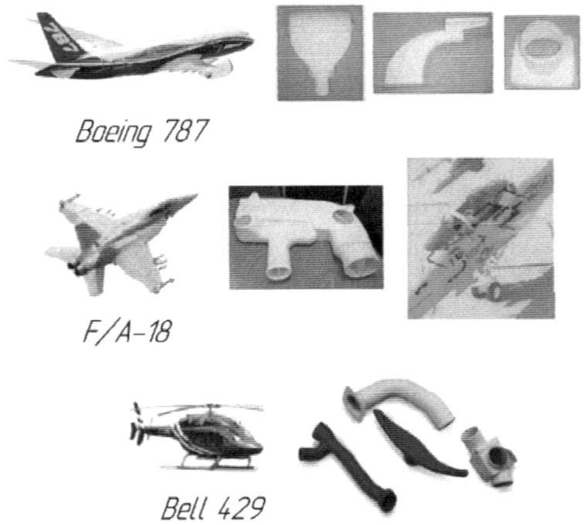

Figure 1. Parts and units of air ducts obtained by the technology of FDM 3D printers [2].

Russian engineers are also actively involved in introducing parts and assemblies obtained using FDM 3D printing into the production of Russian aircraft. Figure 2 shows air ducts for the MC-21 aircraft [3].

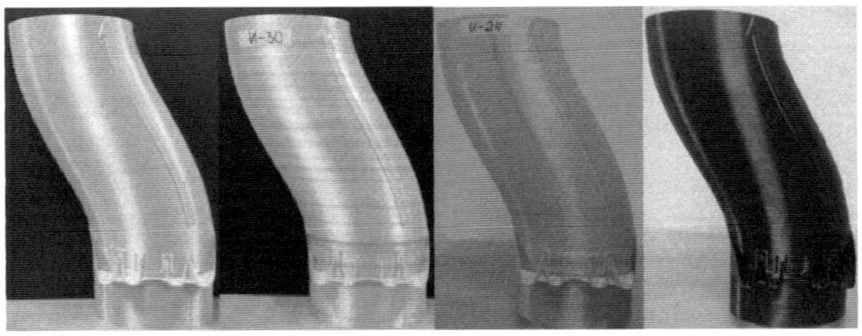

Figure 2. Air ducts obtained by FDM 3D-printers for the MC-21 aircraft [3].

In addition to air ducts, FDM 3D printers are used to manufacture interior parts for aircraft and helicopters.

Unlike aircraft construction in drone construction, using FDM 3D printing, you can make not only individual parts and assemblies, but the entire UAV body. So engineers from the University of Sheffield designed and manufactured the UAV body, completely manufactured using an FDM 3D printer [2, 4]. Figure 3 shows a UAV from the University of Sheffield.

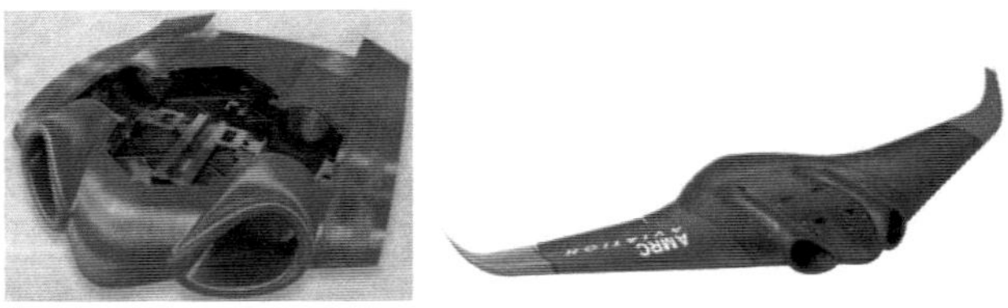

Figure 3. UAV of the University of Sheffield [4].

Every year, UAV and aircraft manufacturing companies are expanding the range of parts and assemblies obtained by FDM 3D printing technology.

3. Classification of defects in parts obtained by FDM 3D printing technology

Based on the analysis of parts obtained by FDM 3D printing technology, a classification of the most arising part defects has been compiled:

a) the first layer of the part during 3D printing does not stick to the desktop;
b) clogged print head nozzle;
c) overheating of the polymer material;
d) the polymer material is not sufficiently extruded;
e) too much polymer material is extruded;
f) the formation of hairlines between parts of the printed part;
g) the formation of breaks in the upper layers of the part;
h) separating the layers of the printed part;
i) small parts of the part are not printed;
j) the formation of waviness on the surfaces of the printed part;
k) tearing inside parts;
l) scratches on the surfaces of the printed part;
m) high surface roughness.

FDM 3D printing also introduces other types of defects. However, they occur less often than the ones listed above and therefore will not be considered in this article.

4. Fault tree

The use of sections to divide the text of the paper is optional and left as a decision for the author. Where the author wishes to divide the paper into sections the formatting shown in table should be used.

A fault tree was built for the functional-structural analysis of the factors causing defects in parts in FDM 3D printing.

Fault tree analysis is a technique for identifying and analyzing factors that may contribute to the occurrence of some undesirable event. The fault tree can be used for qualitative analysis — identification of potential causes and ways of failure, defects, or for quantitative — to calculate the probability of a vertex event in the presence of information about the probabilities of event-factors [5].

In this work, defects of parts produced by FDM 3D printing technology are used as failures. To build a fault tree, it is necessary to draw up a technological process, determine the work items and describe the functions of the work items [6].

The FDM 3D printing technological process consists of the following operations:

1) equipment calibration;
2) preparation of a 3D model for 3D printing;
3) download the file (G-code) to the 3D printer;
4) 3D printing of the part;
5) removing of a part from the working chamber of a 3D printer;
6) post-processing of the printed part (if necessary).

Work items are factors that affect the operations of a process. The working elements of the technological operations of FDM 3D printing are:

1) operator;
2) 3D printer;
3) 3tools (probe, spatula for extracting a part from the working chamber, sanding tool for post-processing);
4) 3D printer software;
5) a slicer program for processing a 3D model and translating it into G-code.

Functions are described for each work item below. Functions are the purpose of a workflow item for performance. A process work item function reflects the input to a process activity for creating process / product functions. These are controllable process parameters that provide the characteristics of the technological operation and products [6].

Figure 4 shows the FDM 3D printing fault tree of UAV parts and assemblies.

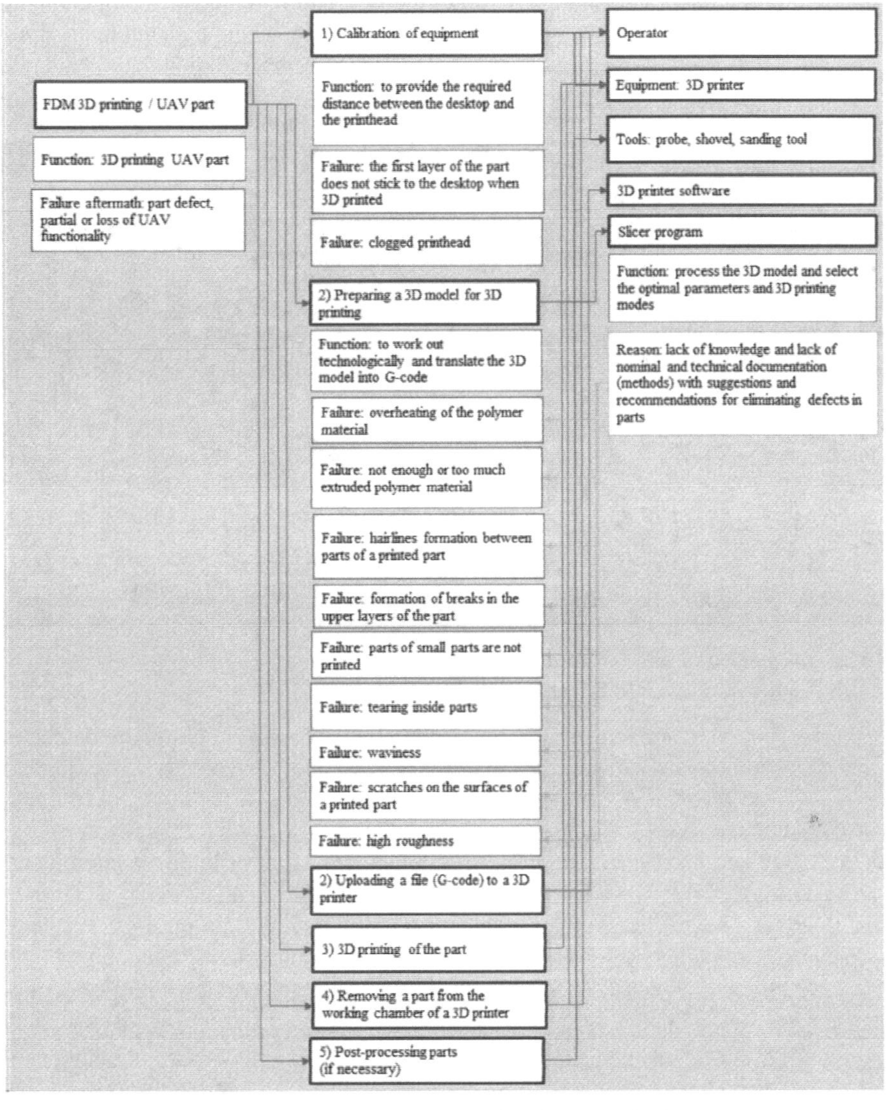

Figure 4. Fragment of the fault tree for the 3D printing process.

As you can see from Figure 4, the operator's work in the slicer program causes the largest number of potential failures. A slicer program is a program for processing a 3D model, selecting technological parameters and 3D printing modes, and then translating a 3D model into machine code (G-code) that a 3D printer understands. The reason for the failures is the lack of knowledge and experience (which is especially typical for novice operators) and the lack of normative and technical documentation (methodology) with proposals and recommendations for eliminating (failures) defects. Calibration of the 3D printer desktop also plays an important role.

5. Methodology for improving the quality of FDM 3D printing of parts and assemblies of UAVs and aircraft due to parametric software changes

Using the constructed fault tree for the 3D printing process, a method for improving the quality of FDM 3D printing was developed, based on parametric changes in the slicer program and 3D printer software. The basis of the technique is a table in the rows of which the types of defects are listed, and the columns list the possible causes of defects and the ways to solve them using changes in the 3D printer software code, machine code (G-code) and parametric changes in the slicer program.

Figure 5 shows a fragment of the table from the developed methodology.

No	Defect type	Cause of the defect	Solution using 3D printer SOFTWARE	Solution by changing the G-code	Solution using a slicer program
1	First layer of the part does not stick to the 3D printer desktop	1) overlay the first layer with high print speed; 2) incorrect table heating setting; 3) incorrect height of the first layer; 4) the desktop is not aligned.	1. #define AUTO_BED_LEVELING_3POINT #define AUTO_BED_LEVELING_LINEAR #define AUTO_BED_LEVELING_BILINEAR #define AUTO_BED_LEVELING_UBL (if there is a sensor) #define MESH_BED_LEVELING (if there is no sensor) 2. #define DEFAULT_MAX_FEEDRATE {500, 500, 2.25, 45} 3. #define BED_MAXTEMP 150	1. M851 Z0 (setting zero Z offset) G28 (Parking, reset with zero Z offset) G29 (calibration after setting the desired calibration mode) 2. G0 FN (Set the travel speed to N mm/min.) 3. M140 SN или M190 SN	1. Lower the print speed 2. Set a different table temperature 3. Change the layer height
2	Polymer material not extruded enough	1) low extrusion ratio; 2) mismatch of the polymer thread diameter.	1. #define DEFAULT_NOMINAL_FILAMENT_DIA 3.00 2. #define PREVENT_LENGTHY_EXTRUDE #define EXTRUDE_MAXLENGTH 200	1. M200 DN (to set the filament diameter N) 2.M220 S80 (Set the feed rate to 80%) G10 S1 (Requires FWRETRACT)	1. Set the thread diameter 2. To increase the value of the flow
3	Hairlines formation	1) the distance of retraction; 2) retraction speed, 3) high temperature of the extruder.	1. #define RETRACT_LENGTH 3 2. #define RETRACT_FEEDRATE 45 3. #define HEATER_0_MINTEMP 5 #define BED_MINTEMP 5 #define HEATER_0_MAXTEMP 285 #define BED_MAXTEMP 130	1. G1 Ennn (Set the retraction value nnn mm) 2. G0 Fnnn (Set the feed rate to nnn mm/min) 3. M104 или M109	1. Set the "retraction distance"parameter) 2. Set the "retraction speed"parameter)

Figure 5. Fragment of a table from the methodology for improving the quality of FDM 3D printing.

The advantage of the developed methodology is the ability to quickly eliminate defects in parts and components without interfering with the design of the 3D printer.

6. Conclusions and discussion

Thus, a methodology has been developed that will allow operators of additive equipment to quickly find and eliminate defects in FDM 3D printing and obtain high-quality products. The technique is mainly aimed at helping novice operators who do not have the knowledge and experience in eliminating defects in parts and assemblies obtained by FDM 3D printing technologies. It should be noted that this technique does not take into account design imperfections and failures of components and mechanisms of 3D printers. Further research will be aimed at finding dependencies between defects and nodes and mechanisms of additive equipment and at improving the design using modern quality tools and robust design [7–10].

Acknowledgments

The authors express their gratitude for valuable advice in planning the study and scientific advice to Doctor of Technical Sciences, Professor Antipov Dmitry Vyacheslavovich and Candidate of Technical Sciences, Associate Professor Dmitriev Alexander Yakovlevich of the Department of Aircraft Production and Quality Management in Mechanical Engineering, Samara National Research University.

References

[1] Dmitriev A Ya, Zagidullin R S and Mitroshkina T A 2019 *IOP Conf. Series: Materials Science and Engineering* **714** 012006

[2] Malfitano B 2018 3D printing in the aircraft industry *Additive technologies* (*Electronic Materials*)
[3] "Irkut" corporation 2020 3D printing elements structures of aircraft on-board systems from thermoplastic materials *The Union of Aviation Industrialists* (*Electronic Materials*)
[4] University of Sheffield Advanced Manufacturing Research Centre 2020 FDM-printed fixed wing UAV (*Electronic Materials*)
[5] Melnikov A Yu and Solomko Yu A 2016 *Science management: theory and practice* (Kiev: The Institute of the Economy of Industry of the NAS of Ukraine) pp 166–172
[6] Antipov D V and Klentak A S 2019 *Proceedings of Samara Scientific Center of the Russian Academy of Sciences* vol 22 (Samara: Federal State Budgetary Institution of Science Samara Federal Research Scientific Center of Russian Academy of Sciences) pp 11–15
[7] Dmitriev A Ya, Mitroshkina T A and Vashukov Yu A 2016 *Robust design and technological preparation for the production of aircraft products* (Samara: SSAU) p 76
[8] Zagidullin R S, Barinov P V, Burkova V A, Glushkov S V and Mitroshkina T A 2019 *Quality and life* **2** (22) (Moscow: Academy for quality)
[9] Dmitriev A and Mitroshkina T 2019 *IOP Conf. Ser.: Mater. Sci. Eng.* **476** 012009
[10] Flores I, Coatanea E, Salmi M and Tuomi J 2014 *1st International Symposium on Robust Design* (*Electronic Materials*) pp 135–145

AUTHOR INDEX

Abykanova, B T ... 146
Beyseev, S A ... 70
Demidova, L A ... 1, 10
Dmitriev, A Ya ... 17, 178
Efremov, A V ... 23
Emelyanova, O ... 46
Fevralskikh, A V ... 29
Gazanova, N Sh ... 103
Golovnev, I G ... 36
Grumondz, V T ... 41
Jatsun, S ... 46
Kenzhekhanova, M B ... 131
Kharkov, V P ... 53
Klesareva, M ... 118
Kondratov, A P ... 59
Korolev, E I ... 103
Korsun, O N ... 53, 63, 70
Kovshov, E E ... 76
Krasavin, I V ... 84
Kulabukhov, Vladimir ... 91
Kurbanismailov, Z M ... 165
Kuvshinnikov, V S ... 76
Lamtyugina, A V ... 135
Lapshin, K V ... 36
Lazarev, S A ... 103
Leon, A S Martinez ... 46
Luchkov, A N ... 97
Lukashev, P E ... 103
Lushnikov, B ... 46
Melnichuk, A V ... 115
Mitrofanov, O ... 118, 124
Mitroshkina, T A ... 17
Morocho, L M Mosquera ... 46
Morozov, N I ... 41
Mukhachev, P A ... 36
Muzhichek, S M ... 131
Napreenko, K S ... 135
Naukenova, A S ... 70, 186
Nesterov, V A ... 115
Nikolaev, S V ... 63
Nikulchev, E ... 142
Om, Moung Htang ... 41, 53, 63
Osipenko, L E ... 76
Osman, M ... 124
Pavlov, V I ... 131
Pechurin, A ... 46
Petrova, N A ... 10
Pravidlo, M N ... 53
Pronyakin, V I ... 146
Prutzkow, A ... 151
Ragutkin, A V ... 103
Rodionov, N V ... 190
Rogachev, G N ... 158
Rogachev, N G ... 158
Russkikh, I V ... 17
Sarmiento, C A Nolivos ... 46
Sebryakov, G G ... 131
Sidorov, M I ... 103
Skrypka, V L ... 146
Sokolov, O V ... 36
Sokolova, Yu S ... 1
Stavrovskiy, M E ... 103
Sudakov, V A ... 115
Syltanova, Y M ... 59
Sypalo, K I ... 115
Tarlanov, A T ... 165
Tiaglik, M S ... 23
Tolstych, S V ... 131
Tsatsin, A A ... 175
Tulekbaeva, A K ... 70, 84
Uali, A B ... 186
Voronka, T V ... 23
Vyshinsky, V V ... 36
Vysotskaya, M V ... 178
Yashin, A V ... 84
Yurko, V N ... 70, 186
Zagidullin, R S ... 190
Zezin, N I ... 190
Zhuravleva, G N ... 59